ENERGY

WHAT ABOUT IT?

ENERGY
WHAT ABOUT IT?

Jean Pierre Fillard

University of Montpellier II, France

World Scientific

NEW JERSEY · LONDON · SINGAPORE · BEIJING · SHANGHAI · HONG KONG · TAIPEI · CHENNAI · TOKYO

Published by

World Scientific Publishing Co. Pte. Ltd.

5 Toh Tuck Link, Singapore 596224

USA office: 27 Warren Street, Suite 401-402, Hackensack, NJ 07601

UK office: 57 Shelton Street, Covent Garden, London WC2H 9HE

Library of Congress Cataloging-in-Publication Data
Names: Fillard, J. P., author.
Title: Energy : what about it? / Jean Pierre Fillard, University of Montpellier II, France.
Description: New Jersey : World Scientific, [2024] | Includes bibliographical references.
Identifiers: LCCN 2023023172 | ISBN 9789811267468 (hardcover) |
    ISBN 9789811267475 (ebook for institutions) | ISBN 9789811267482 (ebook for individuals)
Subjects: LCSH: Force and energy. | Power resources. | Science and civilization. |
    Technology and civilization.
Classification: LCC QC73 .F55 2024 | DDC 531/.6--dc23/eng/20230815
LC record available at https://lccn.loc.gov/2023023172

British Library Cataloguing-in-Publication Data
A catalogue record for this book is available from the British Library.

For any available supplementary material, please visit
https://www.worldscientific.com/worldscibooks/10.1142/13174#t=suppl

Desk Editors: Logeshwaran Arumugam/Amanda Yun

Typeset by Stallion Press
Email: enquiries@stallionpress.com

Preface

Some preliminary words are required to introduce this immense subject, which has been approached so many times in the literature, and with very partial developments centered on technical topics. The aim of this book is not to deal with technicalities but rather to give a global and higher (philosophical or universal) point of view which might help readers reach a significant enlightenment.

This subject of Energy is a purely abstract concept, and does not refer to any material objects even if they are at the origin of the idea. To make things much clearer, I would simply say that nobody has ever seen a kilowatt-hour (KWh) even if everyone knows that swinging a big stick can be full of energy! This virtual concept, qualitatively as well as quantitatively, accompanies transitory situations occurring with physical, chemical, biological, radiative or other transformations, which could arise spontaneously or create impact on a material system. This could apply both on a very small scale and on a gigantic galactical scale.

This is an immense subject which, by its tremendous breadth and diversity as well as its extension and human involvements, makes it quite a "saga". The concept of energy has not only been a constant driving force behind our civilizations, but also the source of so many conflicts and wars around the world. It remains a major factor which has guided and continues to guide our destiny for all eternity.

Given its implications across the spectrum, energy has required a variety of reference units for various situations. They are all domain specific but also convertible into each other for convenience. This dynamic spreads over huge scales from "femto-" or less to "peta-" or more.

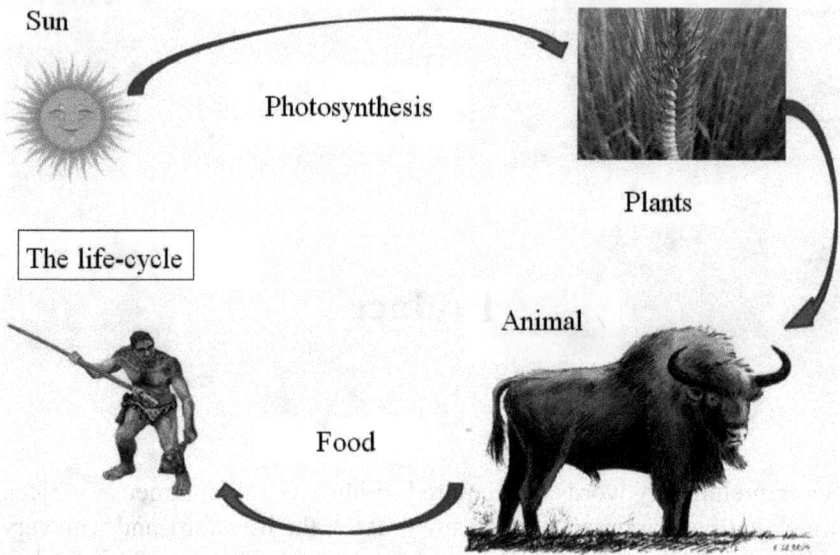

Figure 1.   The life cycle.

The life cycle (Figure 1) always remains fundamentally the same with the energies that accompany us: the Sun (at the origin of most things) constantly showers us with its beams. Photosynthesis in plants assimilates carbon,[1] thus providing food for animals, which in turn become food for humans! Following all that, the exploitation of induced energies (wind, water, wood, petroleum and gas) will do the rest, and in many other ways, contribute to the devolution of successive civilizations.

Surprisingly we can find many books written about energies, but quite nothing about Energy itself. The only extensive treatment was given by Pulitzer Prize winner Richard Rhodes, who described the story of the usage of energy in its different forms from earliest times, events, epochs, places, approaches, examples, disasters as well as triumphs, in his book *Energy: A Human History.*[2]

Our living environment benefits from this concept of energy to bring us comfort, travels, food and so on. This is the result of natural phenomena which are used for practical purposes. We now have reached a point

---

[1]Never forget that CO_2 is an essential contribution to the life cycle.

[2]Rhodes, R. (2018). *Energy: A Human History*. Simon & Schuster.

where we use energy at an unprecedented level to provide us with comfort and improve our life expectancy — consequently resulting in massive population growth (the United Nations believes we have already crossed the 8 billion people mark as of November 15, 2022[3]).

Due to our greed, and uncontrollable quest for more, this idyllic scenario, however, implies some unpleasant inconvenient consequences: pollution, and resource depletion. Nature no longer imposes its law. We now must adapt and manage all these resources. Malthus warned us of this in strong words a long time ago.

It is admittedly unthinkable to regress from the comforts that we have acquired, but we have begun to see the impact of that on the global climate of the planet. Everything in our world comes with a price.

The only latent energy which has survived through the ages comes from matter itself; it remained buried in the very bodies of atoms until the modern times when humans discovered how to harness this plentiful (if not inexhaustible) and so difficult to control, "nuclear" energy.

This book aims to be a free and general reflection and is divided into these sections:

- What Does Energy Mean?
- About the Various Forms of Energy
- Discoveries Follow Each Other in a Precise Order
- The Historical Implementation of Energy with Tools and Machines
- The Pivotal Role of Knowledge
- Civilizations and Induced Issues
- Would Energy be Lasting?
- Where Does Energy Come From?
- Would There Be a Conclusion?

---

[3]Adam, D. (2022). World population hits eight billion — here's how researchers predict it will grow. *Nature News*. Retrieved December 13, 2022, from https://www.nature.com/articles/d41586-022-03720-6.

About the Author

Jean Pierre Fillard was born in (past) French Algeria and retired from the University of Montpellier II, France, as Professor Emeritus in 1998. He is the author or coauthor of over three hundred publications in the field of electronics and optics. During his lifetime he traveled widely in a number of countries all over the word to share his experience. More details can be found on Bing pages.

Contents

Contents

Chapter 1

What Does Energy Mean?

"Energy" (ἐνέργεια/enérgeia) is a very ancient Greek word to mean the idea of an acting force; it was initially proposed by Aristotle and later revived by Thomas Young[1] to stand for a purely virtual concept related to physical phenomenon as well as spirituality. In this first chapter we will present an overview of everything concerning the use of the convenient word "Energy".

It took some centuries before humans began to have a better idea of what Energy could be, first in the basic case of the movement of objects, known as Dynamics.[2] Then came the transformations between heat and "work" (First Principle of Thermodynamics). The following decades have led us to explain heat at an atomic level as an atomic or molecular mechanical agitation in a statistical way.

When Isaac Newton (1642–1727) drafted[3] the laws of movement, he did not care about energy but only about forces, and introduced the idea of mass instead of weight which led to the discovery of *vis insita* (inertia) and *vis impressa* (given movement). This was the first step toward the laws of movement.

[1]Smith, C. (1998). *The Science of Energy — A Cultural History of Energy Physics in Victorian Britain*. The University of Chicago Press.
[2]Balian, R. (2013). La longue élaboration du concept d'énergie. *Consulté le*, 3(09), 2013.
[3]Newton, I. (1687). *Philosophiae Naturalis Principia Mathematica*. Londini, Jussu Societatis Regiæ ac Typis Josephi Streater. Prostat apud plures Bibliopolas.

1. About the origins

Since the dawn of time, Man has striven to improve his quality of life. That created new, greater and ever more binding needs. Through this process, a parameter developed to such a point as to become essential: energy. This external complement was derived from our environment in very diverse ways.

Whenever a new form of energy was discovered and introduced, it led to violent changes in our established ways of life — changes in human civilization so dramatic there was no possible way back.

At the beginning, humans lived an animalistic cycle of life that was rather simple and well established; it may be summed up as "[to] eat, procreate and die". Whether the final intention of Nature (or some say, God) have been, through such a food chain and transformations, to lead the human species to an ultimate goal (I was going to say a "higher level" of existence)? That is another story!

In that earlier period, millennia passed in an utterly precarious state, while the human species survived and even evolved greatly (as explained in Darwin's theory of evolution by natural selection).[4] Nobody had the idea of belonging to a "civilization" although the Sun's energy was already available to help them live. Man remained still an animal, similar to other "hunters and gatherers". As it was for every animal, life was a purely bio-chemical process, no matter its level of intelligence, and there was no attempt to imagine something else outside of what was routine. Energy was a matter of muscular ability, which was all that mattered.

And then, one day, a Cro-Magnon man (to give a name to this forerunner), a bit smarter than the others, found a branch set alight — either by lightning or some wildfire — and dared to bring it back to the cave where he took shelter. After they had overcome their first fears, his companions discovered for the first time that this spontaneous calorific energy would be welcome at home, thus providing light and heat. Then they got accustomed to maintaining a fire, even making war with their neighbors to protect their new property.

Later, they also discovered that meat was tastier after being roasted — all of this on top of keeping away predators which did not dare

---

[4]Darwin, C. (1859). *On the Origin of Species by Means of Natural Selection, or, The Preservation of Favoured Races in the Struggle for Life*. J. Murray.

Figure 1.1. With the energy from fire, the first "civilization" was born.

to approach the cave anymore (Figure 1.1). So, a new way of life suddenly developed for these humans that discovered the magic of fire's energy.

This was the very first true glimmer of intelligence in these thick brains. This was the first step to separate this particular species from the other animals and stimulate the "evolution" dear to Darwin. Man had entered, for the first time, through the door of a "fire civilization" to master this primitive power. The notion of comfort and fulfillment obtained from energy would then lead to the transformation of our civilization to make life easier but also, at the same time, generating some new unexpected, unavoidable obligations (like the collection of dry wood to keep the fire burning).

It was the first time that anyone used an external, non-muscle-based form of energy, and so generated a new need that soon became essential to everybody. Of course, humans of the time had not yet extended this empirical practical discovery to reach the intellectual and virtually new concept of energy, which would take millennia to mature.

Of course, also from that time on, Mrs. Cro-Magnon had to look around for the indispensable dry firewood as Mr. Cro-Magnon was to look for meat (after all, during such a remote epoch, there were no supermarkets).

That was the very beginning of fuel constraints! Up to the present, our quest for energy has remained a permanent concern to fulfill endlessly new and increasing requirements.

Though humanity's awareness, grasp and dependence on energy has come far since our Cro-Magnon days, the concept of Energy (and humanity's understanding of it) has remained a largely virtual one; it does not refer to a precise physical entity or phenomenon, and may be related only to a potential or effective transformation of a material system. This remains a mental concept, so this definition might appear somewhat confusing! All the more so as this concept, initiated from the physical sciences, has been extended to various fields such as philosophy, spirituality and more. An extended view can be found in the concept of "work" which directly implies energy.

Energy is a concept imagined to be responsible for everything in our world that continuously evolves with time. To do anything requires a transformation of the previous state, and a quantitative contribution of this intangible fluid is the means to complete this transformation. To rely on this imaginary fluid on the regular basis involves flows that might be seen as positive or negative, considering the situation. When something new does appear, that means that, in one way or the other, energy has been exchanged.

Einstein said: "Everything is energy and that is all that is to be understood in life." To conclude he invoked the speed of light and the mass of a moving particle, related in the infrangible famous equation $E = mc^2$.

Then, a step further would be to take stock of these exchanges. It looks like mathematics, as an abstract universe, would be a convenient tool to manage energy by following the varying aspects and equivalences it could take and the corresponding physical laws which help to complete the process.

As a principle, energy is considered conservative; that means, if it disappears somewhere then it surges elsewhere in equal proportion, possibly in several forms. In cases where it seems that this law is not respected, we must take a closer look at the problem.

That was the case, for instance, with the β radiation arising from the deactivation of a nucleus (B or N); something was missing in the energy balance, and it took a long time to reach the physical evidence of the "neutrino" emission first suggested by Pauli. This nearly undetectable particle resisted the spotlight for a long time.

2. Energy vs time

Energy cannot be dissociated from the concept of time; these concepts are intimately related. Time is the very parameter which controls the flux of energies, whatever they are. If it were not there, then everything would freeze over, and the implied energies could not be exchanged. Time is also the very parameter to measure the flux of energies during their transfer if we could measure this transaction with confidence.

To get rid of the sun, which was for a long time the only way to measure time, everything which outputs a continuous and steady flux of energy has been monitored to get a measurement of the time elapsed. Time and energy are intimately related concepts. However, historically, the concept of Energy has taken a long time coming to evidence; much has been achieved in sciences (e.g., astronomy) without the need of a formal concept of energy. The sky became the laboratory of the first science due to the star energy implied in astronomy.

Then it became necessary to measure the passing of time using physical observations or phenomena on the regular basis of an observed periodicity. The most ancient and easily accessible "clock" was undoubtedly the moon whose regular movement led to the invention of various calendars following the phases of the moon; however, more precise and versatile means were required and then more recently mechanical and then electronical clocks appeared.

To measure time requires the energy of an fluctuating steady periodical reference (or deemed to be so). For example, the sun was initially used to provide its energy and make a shadow of a stick vertically planted in the ground (the *gnomon* of the Greeks). The shadow regularly rotates and gives an appraisal of the time elapsed during the day when the moon was absent. As a first step, this was a better method than waiting for the moon which is not easily observed during the day.

However, more sophisticated means were proposed over the centuries by talented inventors. Initially, use was made of gravity-driven fluids such as water (*clepsydra*) or sand (hourglass) but those proved inadequate still; then came the pendulum, with the Law of the Pendulum (also known as isochronism) established by Galileo and then used by Huygens thereafter. Then units of time were proposed which accompanied the concept of power, defined as the total energy generated during a unit of time.

Every of these "clocks", even mechanical or sophisticated electronical ones, requires an unavoidable contribution of energy. They only aim at giving an actuated image of that virtual concept of time, which is inevitably ever changing. Energy and time remain indivisible and as well as abstract. However, it very frequently happens that we must wait before expecting energy to be generated; then we must store this energy and only let it go on purpose. Then it becomes a potentiality of action, of physical transformation in a system.

Admittedly again, power or energy, as well as information for instance, are virtual mathematical concepts intended to materialize somewhat intangible and transitory which accompany every physico-chemical transformation in our environment, the externalized balance of a reaction. Following their physical origin, the mathematical formulation can be adapted; that from the dynamic movement of the objects and kinetic energy of Leibniz up to the famous Einstein formula ($E = mc^2$) of the ultimate energy stored in matter or the electron–volt issued from the atomic physics and electronics. The full range of tools today covers every well-known situation with corresponding mathematical symbols to make things related:

$E = \frac{1}{2}\,mv^2$ (Leibniz, Descartes)

$E = mgh$ (Newton)

$E = m{\cdot}m'/d^2$ (Newton too)

$E = JQ$ (Carnot)

$E = h\nu$ (de Broglie, Planck, Einstein)

$E = mc^2$ (Einstein)

$E = eV$ (unknown author!)

This classification will help to trace the path followed by the many transformations which follow one after another from the initial source implemented until the final intended usage. Indeed, Energy always follows a cascade effect to reach the final form we eventually need. Energy remains the source of all things.

Here are some examples of energies which could be exchanged following well established balances:

- Mechanical
- Inertial

- Thermal
- Acoustic
- Radiative
- Electrical
- Electromagnetic
- Chemical
- Nuclear
- Gravitational
- Gravity [Is this doubled? — ECY]
- even Capillarity

They all can behave as "potential" energies.

Of course, the more "natural" energy is obviously the animal and plant energy which grows "spontaneously" under the influence of the sun and proceeds from the assimilation of food in a close cycle. This energy was largely used from ancient times to help Man survive and enable him to achieve his works. That is quite a vicious cycle as it was sketched above. Later after Cro-Magnon Man, animal domestication and agriculture largely improved the cycle, thus giving a new dimension to the human possibilities of realizations with the passing of time.

Energy cannot be obtained from nothing; it is in no way renewable (despite our best hopes) it only changes and dissipates! To acquire energy, we always need a preexisting source of "potential" energy, that is to say available, ready to develop. Then comes the advantage of being able to store energy in one form or another in order to "potentiate" this energy for the purpose of being used later. However, energy tends to escape, and is not easily boxed. Each implementation of a source of energy inevitably leads to unrecoverable yield losses.

This is the case of hydrogen, often considered the ideal fuel. However, there is no natural source providing us with pure hydrogen; we must produce this gas and, in the process, use up a lot of non-recoverable energy to split the tight H_2O molecule. Then the tasks of storing, transporting, and using this inflammable gas proves not to be so simple. Using hydrogen requires very special conditions to be cost-effective. Again, we can juggle with energy, but not create it out of nothing.

A world that lacks time would, as well, lack energy. This is inconceivable. The role of time plays an essential determination in the delivery of energy.

Figure 1.2.   Wasted energy, what a pity!

That does not prevent Nature from demonstrating uses of the power of energy (wind, tides, storms or lightning for instance, among others) without anyone to trigger the show.

However, these spontaneous energy developments scarcely can be recovered (Figure 1.2) for human use. Energy, when spreading out, does not care about humanity; energy is life, and it remains a constitutive part of the Universe. One of the most famous exceptions remain the geothermal source of heat from which some fortunate few (such as Iceland) can benefit, as a stable and valuable contribution to the energy balance of their country.

Earth itself is a fantastic energy tank considering the internal heat maintained in the core at very high temperature and high pressure, dating from the beginning of things. Several masses of liquid lava occasionally find a way to the surface and give rise to violent volcanic eruptions.

We know something about Earth's structure and its corresponding internal physical phenomena but the inner core (Figure 1.3) at the very center still remains a mystery. There lies the energy remaining from the first "post-Big Bang" times; there pressure and temperature are incredibly high. We live right above this furnace!

Figure 1.3. The structure of the Earth.

3. Energy and the discoverers

Energy, even if essential, is not sufficient to make things change and to induce Progress; it also needs famous discoverers to move forward. So, after these preliminary stone ages, we slowly emerged into a period where humanity, developing its intelligence step by step, entered the so-called Anthropocene with a new priority that was as essential as food: the quest for energy. In the Quaternary Period, the Anthropocene is the successor to the Holocene. The limits appear rather fuzzy, and some identify that period as the beginning of the age of degradation of nature by Men and the discoverers. In other words, the modern world surged for more than 10,000 years backwards. Since these times human "civilizations" have governed the world and mastered energy!

After the initial discovery of the power of fire provided by wood, the next step lay in the implementation of an improved combustible fuel some 4,000 years ago: coke allowed higher temperatures to be reached and open the way to metallurgy, first of bronze and then iron, with corresponding tools and weapons for hunting, and of course, for war. Deposits abound, especially in China, coming from vast prehistorical forests: an unlimited sleeping reservoir of latent energy. This source of energy remains commonly used in our present world but resulted in pollution. Energy has now to be "clean" and "ecological", and this has become a new pressing requirement for modern societies.

Each discovery and implementation of a new kind of energy inevitably paved a way to transform our way of life; through innovation of tools and instruments as a means to improve our quality of life. Science underpins these discoveries for us to understand and manage energy in a productive way. The only exception to this rule was the recent invention of the transistor which provided us with a supply of intelligence (even if artificial) from the computer.

The transistor has been the serendipitous key to enter the 2.0 world after John Bardeen, William Shockley and Walter Brattain at Bell Lab in 1948[5] discovered the possibility of guiding electrons in a germanium slab using a very small drop of energy! With these mastered electrons inside matter, technology would then take over and make miracles of the electron stunts; miracles which would turn all our lives upside down. Now transistors have become as small as microbes, such that they cannot be seen without a powerful microscope; however, they are ubiquitous, numbering in the billions and consuming large amounts of electrical energy. Even though on an individual level this is negligible, a full circuit requires serious cooling.

This discovery is by no means a new energy, quite the contrary; but nevertheless it has induced a new era with the present 2.0 world of digital technology. It should be considered a unique exception in human history. The next civilization to come should likely be again triggered by the new energy we seek: the huge and cheap source of energy that could soon emerge from the nuclear fusion reaction of light elements that might be conveniently mastered in a near future. Very large efforts are presently developed all over the planet to harness this new potential on an industrial scale, and finally get cheap and abundant energy.

In the table below are the main steps in the development of new energies along the millennia. In correlation, one can also note the corresponding extension of our lifespans.

What has differentiated human development from animal evolution is that, at the beginning, humanity has understood how to tame animal forces in its favor. That was the very starting step after taming the fire. Agricultural society stretches back over 10,000 years, the industrial one only two centuries, and the present 2.0 digital society merely 20 years.

---

[5]Nobel Prize in Physics 1956.

			Longevity	Estimated World Population
Digital age		AI mastered Energy	90	7,5 B
Nuclear age		Nuclear Energy	65	3 B
Electricity age		Electrical Energy	45	1,5 B
Steam machine		Vapour Energy	35	1 B
Bronze Iron civilisation Coal age		Control of fire Energy	30 ?	4 M
Making fire Wood age		Discovery of Energy	< 30	10 K

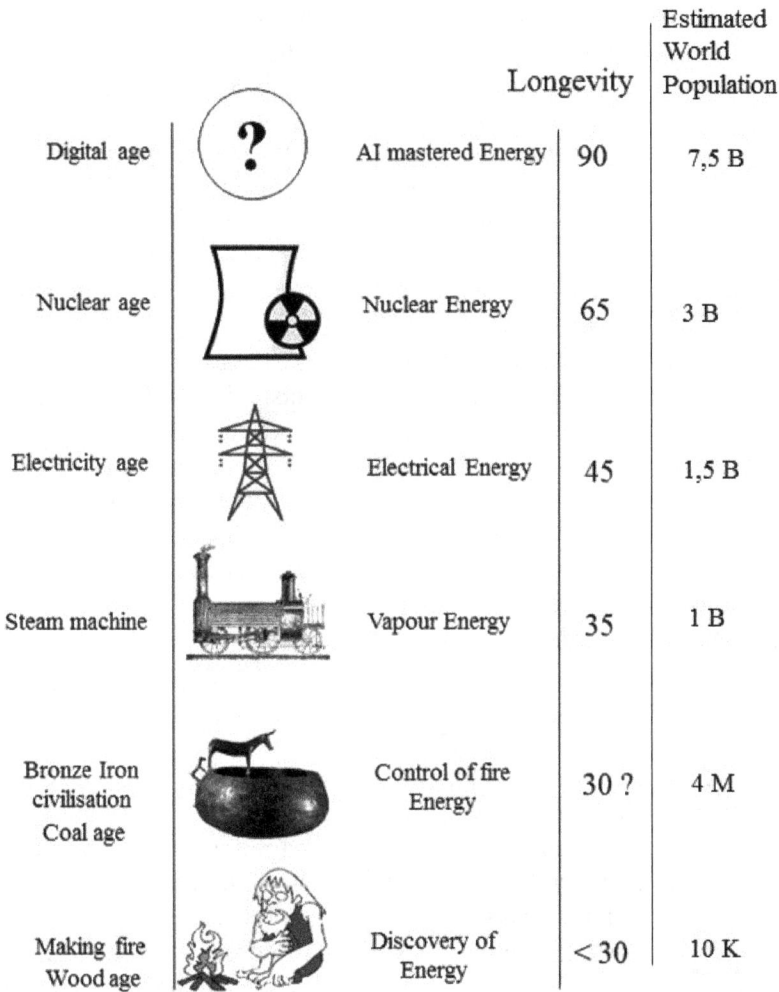

Figure 1.4. Some renowned steps of human civilization.

On a larger scale Ray Kurzweil[6] assembled (Figure 1.4) an infographic of events showing, on a logarithmic scale, the gap between two successive decisive changes of various origins (geological, astronomical,

[6] Kurzweil, R. (2006). *The Singularity is Near: When Humans Transcend Biology*. Penguin Books.

biological or simply human) generated implicitly by an energy implementation, as a function of their age in an inverted scale.

Indeed, this logarithmic plot is astonishingly striking (and somewhat frightening too) if the consistency and the correctness of the data are accepted. This linear alignment clearly demonstrates an hyperbolic trend[7] of the events, starting in the very beginning, close to the Big Bang,[8] with a very wide-spaced evolution but powerful changes along millions of millennia, to reach now an accelerating sequence of steps of weaker amplitude (and weaker energies involved) which pile up rapidly with shorter delays between them.

Let him explain this diagram, reported here below: "the following plot combines fifteen different lists of key events. Since different thinkers assign different dates to the same event, and different lists include similar or overlapping events selected according to different criteria, we see an expected 'thickening' of the trend line due to the noisiness (statistical variance) of this data; the overall trend, however, is very clear".

This plot shows, after the Big Bang at the beginning (on the left side of the graph) when the gaps were in the range of million years, the convergence aims towards very small values in our times, thus showing that less and less energy is required to induce large changes. This should have a hidden meaning. We are now submitted to a real bombardment of new elements that disrupt (or improve?) our living conditions. This is to be attributed to the way we abundantly use various kinds of energy to manage our many different activities. We have even not time to absorb the changes that occur immediately, and this is a unique situation from the beginning of humanity. We have no time to assimilate the change; it just happens, and we must deal with it. What does this mean? I put the question to Kurzweil but he did not answer! He is a guru, not a soothsayer!

More symbolically Vernor Vinge[9] (1995) posited the "Technological Singularity" induced by superhuman intelligence: "within thirty years, we will have the technological means to create a superhuman intelligence. Shortly after, the human era will be ended." This agrees with the

---

[7]This behavior is in no way exponential, as mistakenly said by Kurzweil, but clearly hyperbolical, which is worse still!

[8]Would this Big Bang be considered the Initial Singularity emerging from something before?

[9]Vinge proposed, as a definition that the Singularity would result from "increasing efficiency in the use, processing, transport of matter, energy and information".

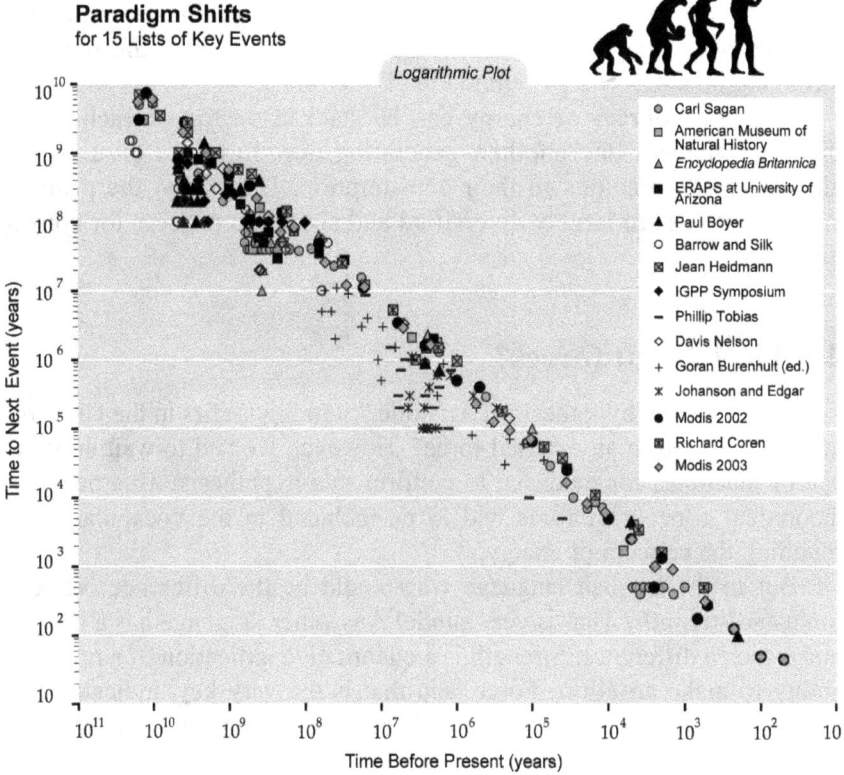

Figure 1.5. The famous Kurzweil plot.

pessimistic view of Francis Fukuyama[10] who said that Singularity is "the most dangerous idea in the world".

All of this conveniently shows that, in the opinion of all these philosophers (and scientists as well), something big (Figure 1.5) is to be expected in the rather near future and this ought to be related with overall Energy.

A consequence of all this is that increasingly comfort of life and medical care, (perhaps intelligence too, at least for some of us) steadily improves and, at the end, the life expectancy increases substantially. All living things require energy, what even the form implied, to evolve and

---

[10]Fukuyama, F. (2003). *Our Posthuman Future: Consequences of the Biotechnology Revolution.* Farrar, Straus and Giroux.

continue to live. Then energy consumption diversifies and increases end-lessly by huge leaps, due first to humanity, then machines and now the many robots which are intended to supplement humanity. At this point we have entered a real energy bulimia in all its forms.

The new sources of energy can be stacked on top of each other; they complement one another, generating new forms of civilization, always more voracious in their consumption. Hopefully, the primary energy sources are largely diversified and simple to process for convenient use.

4. What about forces?

Following the old hypothesis of Aristotle, "nothing is stirs in the Universe unless subjected to an external force". However, we had to wait until the age of machines for evidence to confirm this hypothesis. This new and theoretical concept of force had to be included in the vocabulary surrounding the concept of energy.

But in the English language what could be the difference between force and strength? That is very subtle! Any other language has a need to make such a difference. Strength is a quantitative indication of a potential ability to make an effort. Force, and that is the very key, indicates also a strength but with an added indication of a precise direction or orientation.

A force can be pending or as well moving and then giving rise to energy externally. This is the usual scheme of reasoning in this abstract universe. Force and power are commonly derived concepts devised to accompany the idea of energy in many forms and get a better understanding of what happens.

Force is a concept often associated with muscles, tools, machines or even gravitation; it refers to the way energy could develop and generate a change if the force is allowed to turn in action. Whatever it is, it remains that humans are the only animals who were able to master both animal as well as "non-animal" energy in a profitable way.

5. Beyond physics: Spirituality

The concept of Energy refers to the more abstract and multifaceted concept of science; it only goes back one and a half centuries from the present

after an especially lengthy and winding elaboration with extensive exchanges between the fields of science and technology.

We will present a short overview which can be fruitfully completed by following the monumental work (16 volumes) of C. C. Gillispie.[11] Indeed, Energy is a word of extensive and general use referring to special nonphysical situations related to a behavior or to the inner self.

Dealing with religion, one might evoke the strength of the soul: the intellectual or spiritual energy. This concept was initiated by H. von Helmholtz in 1847.[12] Spirituality is not the exclusive domain of religion, nor a matter of cultural study in the belief systems of primitive people; spirituality is a natural living function of humans where the concept of energy finds its proper place.

The word covers many very different meanings depending on the person, and none may be measurable even if the word energy is largely involved in the discussions. The one who say "to show energy" even it cannot properly be "seen". That concept of "spiritual energy" originating from the physical sciences has been extended to abstract domains like philosophy — as a kind of virtual fluid which has no physical implication — and beyond, even in biology, as suggested by Henri Bergson.[13]

A special mention is to be made of the concept of *qi* (or *chi*) in Chinese which pervades, supports and sustains all living things. Various techniques were formulated to channel this energy to aid in physical recovery. Spiritual energy, positive or negative as well, is better considered as a powerful force which is not limited by an enclosure. Some relate this kind of energy to the influence of celestial bodies in the cosmos. Psychic energy comes from the strength of thought and may induce indirect effects on the bodies. If words have power, this is due to the thought at its origin. Mysticism is a kind of spiritual energy, and the word of energy itself often induces the generation of predictions.

Acupuncture is located between medicine, biology and spirituality of a sort. It aims at restoring a harmonious circulation of nervous vital energy through needles stuck into specific points on the body. Chinese

[11] Gillispie, C. C. (1981). *Dictionary of Scientific Biography*. Scribner's Sons.

[12] Balian, R. (2013). *La longue élaboration du concept d'énergie*. Académie des Sciences. http://www.academiesciences.fr/activite/hds/textes/evol_Balian1.pdf. *Consulté le, 3*(09), 2013.

[13] Bergson, H. (1920). *Mind-energy (L'Énergie spirituelle, 1919)*. McMillan.

philosophy requires a fair balance between *yin* and *yang* through the movement of *qi*, and acupuncture helps to restore a satisfying equilibrium of this vital strength.

6. Nothing is lost, everything is transformed

The concept of work refers directly to an energy transfer that can be considered both ways: as active production or as resistance on the understanding that follows Lavoisier's famous principle[14]: "nothing is lost, nothing is created, everything is transformed".

Everything that lives needs to consume energy to power a transition from one form to another. From a user point of view that also means a "degradation" of the potential energy. Most of the time unwanted heat is produced as a by-product in the process. We only can "consume" an energy already existing. All of that means that, before using energy we have to accept expending more than required for an action.

For Earth, the energy available came from the tireless Sun, and that energy ultimately originated in the Big Bang; some say life is the work of God, a billion years ago, using an optimized energy located in a "primordial soup". Energy gradually spread widely everywhere and became a particular constituent of matter itself after an initial explosion. However, nobody told us when and where that comes from. Who decided about this explosion and where this incredible energy comes from?

6.1. *What do we know about the possible origin of energy?*

This theory of the Initial Energy goes back to a BBC broadcast *The Nature of Things* on March 26, 1949 given by the physicist Fred Hoyle. This infinitesimally tiny universe should have been especially hot, containing an incredible amount of energy reserves. Then it began a perpetual expansion to give birth to galaxies, thus scattering energy that has condensed into matter. Thus was the theory proposed by Einstein.

---

[14]L'énergie, de quoi s'agit-il exactement ?. Jean-Marc Jancovici — Articles et études de Jean-Marc Jancovici. Ce conseiller en organisation propose services et connaissances dans les domaines de l'énergie et du climat. (2011). Retrieved December 9, 2022 from https://jancovici.com/transition-energetique/l-energie-et-nous/lenergie-de-quoi-sagit-il-exactement/.

Cosmic Microwave Background (CMB) radiation was finally discovered in 1964. This low-energy homogeneous microwave radiation at a temperature of only 3 K is considered a remnant of the hot initial epoch due to the abundance of light elements and hydrogen isotopes.

The origin of the galaxies is reportedly attributed to a "gravitational instability" that condenses and stores energy and matter. From that entire genesis still remains the energy we can use on earth today: matter and solar radiation, residues of the Big Bang still within our reach. All our energies come from there. Of course, considered on a galactic scale, our energies used locally on Earth must be estimated as negligible.

The initial explosion continues to push away the galaxies of the first age toward the outer limits of our universe. Some can be observed using powerful telescopes launched in the sky far away, where light pollution from our lights does not disturb the detection of their distant images. Of course, the images we can receive now are millions of millennia old and do not represent an actual current reality. These galaxies continue to expand everywhere, following the energy of their initial impulse.

We are exploring space in ever-increasing ways and continue to discover unexpected phenomena. Our astronomers still await physical evidence of mysterious stars suspected to concentrate a high density of matter (and thus energy); they call this dark matter, which they do know to exist and exert a gravitational influence, similar to "black holes".

7. Energy as the source of all things

In nature, energy is everywhere and arises spontaneously; but in order to use it to deliver the desired outcome, it has to be stored. It is the fuel of our evolution; mastering its use is a task for humans. The starting point was obviously the muscular energy of biological origin, but humanity has moved a step further in learning how to empower itself with external stored energies that, of course, are dissipated into a degraded form mostly through lost heat. Energy is not renewable after being used.

It took centuries to nature to transform plants into wood and then coal (later coke) then fuel and gas. We use all of them as thermal derivatives to generate hot vapor that moves pistons or turbines; each serendipity succeeds the other in a well-established order, the one motivates the other; nothing is improvised. At the end of the cascade lies the nuclear energy deeply immersed in the heart of nuclei.

Meanwhile, non-animal energy has been stored profusely in various kinds of batteries of diverse sizes to animate and run every object we use today: Smartphones, razors, even cars. Not to mention the surgical implants animated by a battery which requires periodically recharging or replacement. This is a main inescapable drawback of those implants.

These batteries must be manufactured starting from materials such as rare earths (lithium, scandium, yttrium, etc.) that are very difficult to mine, and the exhaustion of reserves of each of these elements will pose an serious threat to mankind in the future.

On the other hand, we might also be able to release uncontrolled energies for military purposes such as atomic bombs. Hopefully only for a restricted experiment at present.

Energy provides the means to power and deliver the desired outcomes for us; but without them mankind goes back into the Stone Age. Money is the driving force in the economy that allows the purchase of energy to produce and sustain economic growth.

Our new civilization 2.0 did not provide us with new sources of energy even though its consumption has gone through the roof. Currently the balance sheet totals some 14 trillion watts (10^{13}) of power (at the price of $2 trillion a year) and is expected to double by 2030. The exponential demand of Integrated Circuit production which powers our digital world largely contributes to the growth with Internet and associated "cloud computing" facilities.

In this present civilization, an eminent place has been claimed and established by a newcomer — the Computer. Personal computers (PCs) entered our homes decades ago when big data crunchers created a virtual world we call the Internet. Now we have a global memory which multiplies the performances of our little brains. Everything in our lives now travels through this medium and evidently this traffic, at a global scale, requires tremendous amounts of energy, to such an extent that the corresponding huge data centers must be located close to major electrical power plants to limit the amount of energy lost in the transfer.

However, we may recall that our brain, as one of our key organs, also consumes energy in the body; is that really essential to stimulate intelligence? Would an increased "super-intelligence" require substantially boosting the blood flow and augmenting the oxygenation yield[15] to keep pace with the needs of the brain? Would we have to keep us wondering

---

[15] Would that deserve a comment from Kurzweil?

that our brain is not alone: 8 billion similar brains are permanently similarly burning their energy biologically transferred?

8. The philosophy of the future

This theme of energy has now begun to excite thinkers who actively develop ideas to shape our lives. Among them, the most energetic forecasters grouped together under the banner of an audacious philosophy which they called "Transhumanism"[16] and evidently they are worried about the implications of Energy in our future. These gurus do not belong to any kind of sect; they only work with ideas, develop methodologies, make speeches and talk in workshops and conferences proposing new perspectives to economic or industrial executives of all disciplines in order to anticipate future opportunities of the world economy. As a matter of fact, Energy is a key problem to be considered in this field. Among them, one of the most prominent figureheads is assuredly Ray Kurzweil.[17]

The basis of the transhumanist philosophy relies on the hypothesis of an uninterrupted acceleration of technological means induced by the scientific progress stimulated by an increasing access to energy sources. This point is not explicitly evoked in the transhumanist's discussion but remains in the background as a fundamental requirement. Robots, for instance contribute more and more in our activities and they are greedy for electricity!

This philosophy claims that we are heading imminently toward what they call a "Singularity", that is to say a major breakthrough. This time, we are desperately searching for a new energy source to give to 2.0 his very dimension in the post-humanistic world, especially if a Singularity is to occur someday. Augmented intelligence (artificial or not) is of no use if it does not fit with the corresponding necessary power to act.

What could be the impact of the discovery of an unexpected new energy source on the outbreak of a Singularity? Would that not be a prerequisite? Till now our understanding of the world surrounding us has

---

[16]Fillard, J. P. (2020). *Transhumanism: A Realistic Future?*. World Scientific Publishing Company.
[17]Kurzweil, R. (2006). *The Singularity is Near: When Humans Transcend Biology*. Penguin Books.

improved noticeably, thanks to the physicists and other researchers. However, compared to what remains to be understood, some would think we stay in the Stone Age due to how many mysteries we still have left to decipher. For instance, let us consider the fantastic gravitational energy expended by the Moon to stir up the tides all over the planet. Would not it be possible to benefit more directly from its energy?

Space is sometimes mentioned by transhumanists as a way to promote "improved" humans in a drastically different environment (Mars or even Pluto are on the list of possible planets worth "colonization"). In this context Energy would be the first critical requirement.

Singularity, in some way, means a threshold, a discontinuity, in the collective behavior of humanity. In the past, all the observed "singularities" effectively came from the discovery of a new kind of energy which paved the way to new technological opportunities; afterwards, these opportunities brought a wave of benefit to mankind, as a consequence.

In the special case of a potential unique Singularity to come in our future (as emphasized by Kurzweil), things would not be the same. The threshold would likely arise from a burst in science, in the domains of biology (health, aging, or genetics) and computers (Big Data and Artificial Intelligence (AI), deep learning). Of course, Energy would still be considered in various ways; but rather as a result of the changes, not a direct consequence as was the case before.

To implement energy, it is mandatory to identify a preliminary "potential" source that would be available and ready to be used; but Energy is fugitive, and it is not easy to keep it in a box! Energy cannot be obtained freely from nothing. Energy is a valuable commodity.

Some also think that "Technological Singularities" would result from a burst in Intelligence (assisted by computers?) which would help to realize new serendipities along the prominent contributing role of energy. Some transhumanist philosophers even discussed these opportunities and their applications. Vernor Vinge was the first to coin the term Technological Singularity in 1993 to strictly specify the technological origin of the change expected. The idea of the Singularity holds a powerful intellectual attraction to imaginations and deserves a critical examination (after M. More).

Many authors have fitted human population to a similar hyperbolic growth curve where population would reach infinity at some finite time t. Technology grows proportionally with population; but what about energy?

9. History of the development of the various forms of the energies we use

After fire was mastered with the convenience of coals, higher temperatures were reached, allowing us to melt metallic ores to produce copper or even iron ingots and then tools. For the first time, men were equipped with a considerable new power to amplify natural biological forces. They were then able to cross from the Bronze Age into the Iron Age. Metallurgy was born by inducing an irreversible use of every kind of external energy sources. Every kind of fossil fuel was later assisted by electricity as an intermediate power to make electric arc welding and thus, precisely, to make pipelines to transport the crude oil. So, the loop is closed.

Then it has been possible to harness floods of energies stored in underground for millions of millennia, while waiting for the still more powerful energy made possible by the nuclear reactions. We shall develop this point in particular later more extensively. In the meantime, humanity is striving for its increasingly growing needs; needs that could be fulfilled by the nuclear fusion reaction of light atoms. Let's keep our fingers crossed!

To go back a little bit, primitive tools gave rise to machines that could work independently of any human contribution (except to provide the energy required). Then machines become assembled and integrated into more complex systems to deliver (eventually with the help of robots) an outcome through a cycle of fabrication and distribution. All of that increasingly consumes more energy, to the point that, now, people worry about an induced global warming and the consequences of climate changes, besides the unavoidable pollution close to the cities. CO_2 gas is accused of being at the heart of all these plagues; nobody remembers that this same CO_2 is by no means a poison but the basic food of every vegetable! China and Germany are accused to be the irreducible producers of CO_2 because of their many thermal power stations burning tons of coal mineral extracted from their old deposits still in use today.

Meanwhile, Transhumanist philosophers have developed a principle of "Extropy" (not to be opposed to the concept of Thermodynamical Entropy) defined as "the extent of a living or organizational system's intelligence, functional order, vitality, and capacity and drive for improvement". That's obvious Dr. Watson! Would that help clarify the issue of Energy?

A solution advocated by Kurzweil to contribute to our increasing quest for energy (a need made more acute by an extended longevity) consists of "moving away from the concentrated and centralized installations on which we now depend ... using micro-electro-mechanical systems (MEMS) technology.[18] These devices are manufactured in the form of microchips with an energy-to-size ratio significantly exceeding that of conventional technology. Nano-engineered solar panels will be able to meet our energy needs in a distributed, renewable, and clean fashion. Ultimately technology along these lines would power everything from cellphones to cars and homes. These types of decentralized energy technologies would not be subject to disaster or disruption."

Would this be the "magic bullet" of the Transhumanists? At any rate, it is important to recognize that most (if not to say, all) energy sources today represent solar power in one form or another, even dating back to the earliest days of the Earth itself.

---

[18]Integrated fuel cells technology. Micro-Electro-Mechanical-Systems.

Chapter 2

About the Various Forms of Energy

Several billion years ago, Earth came into being but still was a "hot soup" where carbon-based molecules became more and more assembled and intricate until complex aggregations of molecules, at the surface of the bowl, eventually formed self-replicating mechanisms, and life originated. All of that starting from the available large ambient density of combined thermal, electrical, radiative, and nuclear energies. Ultimately biological systems, like the first cell to be called the last universal common ancestor (LUCA), evolved a precise digital mechanism (DNA) to store information describing a larger society of molecules. That was the first epoch of a new world using all kinds of energies, in a random process, to get an organized reproducible future. The diversity of these energies was the miracle recipe of the success and, hopefully, this diversity still persists today.

From these remote times Man[1] emerged and developed to its advantage the opportunities offered by such a rich range of possibilities (the list is by no means exhaustive).

1. Basic biology

To begin with, let us start with the brain and its neurons which are at the origin of any intelligence. A resting man dissipates a total power of 70 watts in heat (mainly in the brain) that is provided through food. He can generate a maximum of 100 KWh during a year. All of that energy

---

[1] It is the custom to say "man", understanding that women are included in the scope.

come from the cells[2] (numbering some 4×10^{13}), but the population of bacteria should be ten times more numerous and they are as well implied in the process of assimilation of food.

These cells can include three different categories: stem cells (pluripotent), bone cells and blood cells (transporting oxygen throughout the body). Special attention should be paid to the special case of the neurons. These cells located mostly in the brain are highly interconnected in a dense network; they assume a critical role in the management of the body and also a more subtle implication in intelligence.

The brain is engaged all through life, night and day. It manages the five senses, the subconscious (very important) and the body, this is the reason for its high consumption of oxygen, a condition which is directly related to the maintenance of life.

All activities of the body require energy, which is built up through consumption of food that comprises three main nutrition groups (in a sense, sources of energy): carbohydrates (fuel source), protein (muscles, tissue repair) and fats (reserves). This is the bio-chemical process of metabolism active within each cell whose membranes support the passage of these molecules through transformation of energy. The whole machinery of the body obeys the oxygen supply carried by the blood red cells. Oxygen is the fuel of the cells.

From an electrical point of view, electrical current in physiology consists of streams of ions (that means atoms which have lost or acquired an excess electron on their outermost layer). So, these ions as well as electrons carry an electrical charge and are able to follow their paths through the surrounding electrical potential opportunities.

The network of nerves in the body enables the transmission of the brain's commands to the corresponding muscles as stated by Alessandro Volta who stimulated the leg muscles of a frog with an external electrical source made of two disks of different metals separated by a blotting paper drenched with brine.[3] In 1800 he started a dispute with Luigi Galvani who independently discovered what he called "animal electricity". Volta is thus

---

[2]Bianconi, E., Piovesan, A., Facchin, F., Beraudi, A., Casadei, R., Frabetti, F., Vitale, L., Pelleri, M. C., Tassani, S., Piva, F., Perez-Amodio, S., Strippoli, P., & Canaider, S. (2013). An estimation of the number of cells in the human body. *Annals of Human Biology*, *40*(6), 463–471. URL: https://doi.org/10.3109/03014460.2013.807878.

[3]It was actually the first pile demonstrating the "voltaic" effect.

credited with the origin of the unit (volt, V) of the electrical "voltage" and also with the discovery of water electrolysis.

The neuron encodes and transmits information through sequences of output electrical spikes when membrane depolarization reaches a threshold voltage. Identifying the details of the process of how the transfer of metabolic energy takes place is not a straightforward matter, but is key to understand the mechanism of the brain and the way it functions. Energy and its distribution through the neuron are the fuel of brain activity.

We think, thanks to energy. The complexity of the electrical exchanges, inside a neuron and with each other, could be clearly demonstrated (Figure 2.1) in the following scheme.[4] No need to enter here into the details. The energy plant looks quite complex but it works.

All the biology of the cells proceeds from a single initial cell we call LUCA that emerged from the "primordial soup" some 3.5 billion years

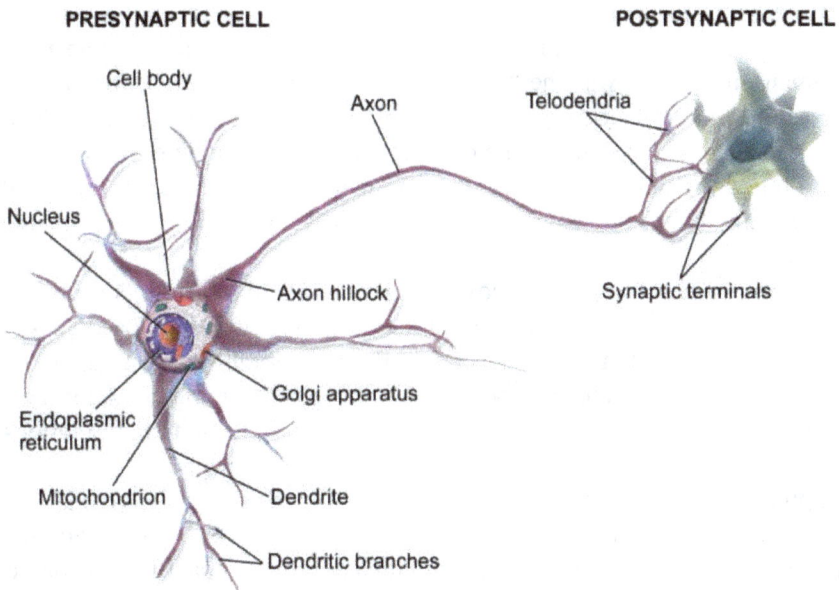

Figure 2.1. Neuron complex organization to exchange energy.

---

[4]Neurons: Where does their electricity come from? Medical Science Navigator (2017). Retrieved December 9, 2022 from https://www.medicalsciencenavigator.com/neurons-where-does-their-electricity-come-from/.

ago and is referred to as the first step before living organisms, which might have originated from viruses. Some utilized a combination of poly-phosphates for storing and transmitting the energy of the ambient medium (e.g., close to hydrothermal vents) for the benefit of this progenitor.[5] This could be the very origin of the bacteria. So this is why Energy (and time) pave the ways to build the biological world.

2. Fire and heat

Fire originates in an exothermal reaction of dry combustible solid or liquid, as well as gas, with oxygen and a heat source. That reaction requires ignition by an external activation energy (heat). This process results in a hot gas emission (CO_2) likely with light. This is one of the simplest ways to start a large emission of heat that sustains the process if there is enough combustible materials and oxygen. As said before, this discovery was at the origin of the development of humanity.

This happened at a time when the only available energy came from human muscles! It was the "Stone Age", where stones, sticks or the like were the first tools to help man achieve a result. After the age of tools, it took millennia to reach the age of machines: the available energy no longer relied on hands but had been transferred to machines: artificial arrangements of the emerging technological innovations such as the wheel.

Later after, it was also discovered that fire could transmit its energy to water, producing a vapor that could increase pressure in a closed vessel, thus providing a stored energy that could be profitably used to again be transformed into mechanical power and perhaps something else later. This is the best example to show the necessary cascade of transformations required to reach a useful adaptation to human needs. This cascade implies a lowering of the global yield for the final energy delivered.

After several unsuccessful attempts, Denis Papin (1679) assembled the first boiler to cook meats, and then had the idea of a piston to transform heat into a mechanical movement; the piston subsequently entered into practical use (Figure 2.2) within pumps in the coal mines.

---

[5]Weiss, M., Sousa, F., Mrnjavac, N., *et al.* (2016). The physiology and habitat of the last universal common ancestor. *Nature Microbiology*, *1*(9), 1–8. https://doi.org/10.1038/nmicrobiol.2016.116.

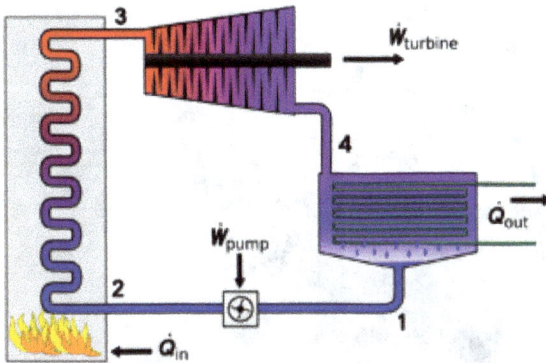

Figure 2.2. Steam turbine scheme.

Then James Watt proposed a real motor powered by steam and simple (or double) action on a piston; this was the very beginning of the "steam machine civilization" extrapolated from the fire discovery.

The thermodynamical process obeys the famous Joule's Law $W = JQ$ which expresses the equivalence of heat energy with corresponding mechanical energy. The step forward was provided by steam turbines giving a better yield as they were designed according to what is called the Rankine (Carnot) cycle. The setup is reported here below.

Note that this evolution has taken millennia (Figure 2.3) to occur. Then locomotives, steamers, powerful cranes, electrical power stations and so on flourished, contributing to a large and distributed energy contribution to the human comfort and facilities. That also gave a fantastic push forward to economy, agriculture, technology and even science. Energy became a catalyst for the first Industrial Revolution. Each improvement generated new applications that, in turn, led to new ideas of development, thus creating a positive feedback cycle as long as scientists gained deeper understanding of the underlying mechanisms.

Little is known of the beginning of the fire story except that *Homo erectus* might have used it some 1,500,000 years before us, but he should have discovered how to light it some 700,000 years before us. It must be said that, at that time, intelligence, previously restricted, began to grow in these simple brains: brain volume began to increase (some think that intelligence is related to the brain volume!).

Nobody actually knows where this fundamental discovery happened: bushland, forests, mountains (storms), cold or warm countries; perhaps

Figure 2.3.   The starting point.

several places before being globally accepted; this is still shrouded in mystery. Some mentioned a cave in Swartkrans, South Africa. The evolution was to learn how to make fire apart from an existing source, like lighting a torch from a fireplace, instead starting fresh fires at one's own discretion with flint stones or rotating sticks made from inflammable materials.

Then this essential discovery may have migrated northward through the Fertile Crescent, toward Europe. Little is said about the distant Oriental world (China, India) but the same process likely developed in the same period. The oldest fossil skulls were found at Hubei (China) and are 936,000 years old and *Homo sapiens* here dates some 110,000 years before the current age. Some innovating Chinese ceramics were found, dating back some 15,000 years.

Then a new important step in energy control came with the observation that coal pieces burned as well as wood and allowed us to reach higher temperatures; that opened the opportunity to melt metallic ores, and these metallurgies gave rise to a tremendous diversity of new tools incomparably more useful than the simple albeit carefully carved stones of hitherto.

Figure 2.4. Possible path for the fire discovery.

The ignitable property of coal was discovered by chance, in a place (Figure 2.4) where a deposit lay close to the surface. This serendipity was an impetus to spelunk more deeply and opened a phenomenon of mining for the subsequent centuries. Energy was there, within reach! Time passed and ways of using coal have evolved. Since that start a still more refined material with a higher temperature potential has been obtained: coke.

The torch blazed ever brighter as it was relayed to an even more potent source of energy: gas. The age of coal was over. Gas had appeared as a newer, practical, and powerful source of energy more amenable to still other more advanced applications: cars, planes and so on. The "black gold" has been tracked down all over the world. In all cases, mastering and improving a new heat source of energy has transformed our societies and our ways of life.

That was the story of wood and coal fire. Later on, extraction of gas increased the thermal yield. But the Industrial Revolution unlocked a whole new energy resource: fossil fuels. Fossil energy has been a fundamental driver of the technological, social, economic and developmental progress which has followed. Now we are dealing with the shale gas which is at the top of the energy sources. The energy story travels a very long way!

However, chemically speaking, the best fuel to consider is obviously hydrogen which is a gas at normal pressure and temperature. Its oxidation reaction with oxygen restores the energy necessary to extract it from the water molecule. Of course it is highly inflammable and requires careful handling so as to avoid explosions or fires.

Usually it can be stored as a liquid at very low temperature (4 K) or under high pressure which is a quite more delicate consideration compared to other usual fuels.

Obviously there is no natural H_2 deposit and this molecule must be obtained chemically (i.e., we must expend energy first). A more obvious solution lies in the electrolysis of water separating H_2 from O, but this operation is very energy-consuming. However this electrolysis can be operated in a stop-and-go process without any inconvenience, and the gas obtained may be stored at leisure.

That property makes H_2 a very attractive solution in comparison to intermittent sources of electricity such as wind turbines or solar converters. Another method of H_2 generation arises from biomass, green waste and bacteria, and techniques for this solution are actively being explored even if an industrial solution has yet to become available.

Hydrogen is now considered as a virtuous approach for energy generation (even if the final energy balance is zero) because of its clean reaction without any CO_2 emission. However, that implies a full conversion of the current gas engine technology and great efforts are being expended to reach this goal. Cars, trucks, trains and even planes are now able to operate safely with that new fuel.

The last step, and possibly the most decisive for the future of this domain of heat generation, relies on the nuclear energy which is the latest technological innovation: fission of Uranium nuclei in the PWR[6] and

---

[6]Pressured water reactor.

perhaps later fusion of light elements in a Tokamak.[7] These may provide us with heat and the derived energies that our new world requires. Would that be enough?

Heat remains the basic origin of all energy sources if we exclude natural mechanical ones such as wind, tides, waterfalls, dams and of course the sun.

In 1856, Rudolf Clausius, referring to closed systems, in which transfers of matter do not occur, defined the *second fundamental theorem* (the Second Law of Thermodynamics) in the mechanical theory of heat (thermodynamics): "if two transformations which, without necessitating any other permanent change, can mutually replace one another, be called equivalent, then the generations of the quantity of heat Q from work at the temperature T, has the *equivalence-value*: Q/T.

In 1865, Clausius came to define the entropy symbolized by S, such that, due to the supply of the amount of heat Q at temperature T, the entropy of the system is increased by $\Delta S = Q/T$. Defining heat as an energy transfer without work being done, there are changes of entropy in both the surroundings which lose heat and the system which gains it. All of this leads to the science of thermodynamics which regulates the laws of interaction between heat and work, two interchangeable measures of energy. In kinetic theory, heat is explained in terms of the microscopic motions and interactions of constituent particles, such as electrons, atoms, and molecules. Heat transfer arises from temperature gradients or differences, through the diffuse exchange of microscopic kinetic and potential particle energy, via particle collisions and other interactions.

3. Electricity and energy

Benjamin Franklin discovered that the energy of the thunderbolt originates in electricity; however, he was never able to use it profitably. So, where does electricity come from and why is it so much appreciated as an outstanding, powerful and versatile source of energy?

We will have to dive into the universe of atoms that basically are made of a nucleus (protons + neutrons stuck together) carrying a positive charge that attracts surrounding negative electrons to achieve global neutrality. All of that results from electrical forces. But inside the nucleus,

---

[7]Russian: *toroïdalnaïa kameras magnitnymi katushkami.*

other proximity forces hold together the nucleons to make a stable structure and attract the corresponding electrons all around. This electrical
charge of the electrons is perfectly conservative whatever its speed, in the
theory of relativity. Here too, the concept of electrical charge is absolutely
virtual, to give an idea on the origin of that particular force.

The coherence of molecules and, more generally, matter comes from
the various electrical influences of the atoms between them. These electrical charges may be mobile in a conducting medium and so constitute an
electrical current able to transfer energy away. The ultimate force results
from the excess or the lack of electrons attracting or repulsing following
the situation. That force exerted between two charges is reciprocally
dependent on the square of the distance; this is governed by what we call
the laws of electrostatics.

But another, unexpected, phenomenon is to appear accompanying the
displacement of electrical charges in a wire: the electromagnetic field (or
Lorentz force) generated all around the wire. The abstract language of mathematics is required to give it a formulation whereas the evidence of it was
purely experimental (Hans Christian Oersted, 1820). Oersted's discovery
also represented a major step forward to a unified concept of energy in electricity and then it influenced the French physicist André-Marie Ampère's
developments of a single mathematical form to represent the magnetic
forces between current-carrying conductors. This new field of forces was
called electro-magnetic induction and is tentatively schemed here below.

The vectorial field $\vec{B}$ is wrapped around any wire carrying a current
as suggested below. Its action appears on any electrical charge q moving
nearby with a speed $\vec{v}$ in the vicinity,

$$\vec{F} = \overrightarrow{qv} \wedge \vec{B}$$

following a 3D scheme depicted by the three fingers of the right hand
(Figure 2.5).

This effect can be amplified in a solenoid winding which is at the
origin of transformers, magnets, electrical motors and so on. Energy is
available there.

Electricity transforms its latent energy in mechanical movement.
Electricity is able to be transferred over long distances without any transfer of matter. Electricity matches with every requirement of energy that
pave the way for its very wide use.

On top of that, modulating an electrical current induces electromagnetic waves that can be used in a variety of applications and carry another

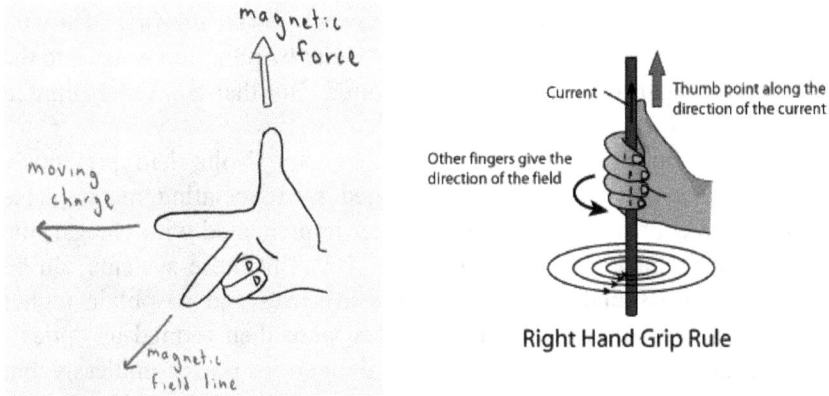

Figure 2.5. A depiction of the right-hand rule on the left, and then the right-hand grip rule on the right.

kind of intangible virtual "matter": information that also requires energy to exist and be transferred.

Radio or TV images can now be readily transferred from place to place without requiring any material link; At a higher level, the Internet arises with its global scale of information storage which consumes huge amounts of megawatt-hours (MWh) day and night with its data centers.

Another important domain where Energy cannot be ignored is chemistry. Molecules require energy to be assembled and chemical reactions occur via the exchange of electrons and then energy, following their electronic affinity.

3.1. *Possible sources of electrical energy*

Electrical energy can be obtained in very large amounts from diverse sources.

- The most widespread sources come from electromagnetic conversion of a mechanical power such as waterfalls, dams, hot vapor plants, wind turbines and so on. This energy can be delivered as direct current or more generally as alternating current. But the major issue of storage remains, in order for the energy to be released sometime later, or somewhere else after being transported to a place where it is to be used. Some applications are more sophisticated such as space energy coming from solar cells or adapted small nuclear reactors.

- The ways and the means to store electrical power: the first idea with water turbines is to reverse the flow of water by pumping water into the reservoir when this power is not required. But that is a rather limited solution.
- Another opportunity comes from chemistry. Volta had previously shown that electricity can be obtained by associating metal plates (zinc and copper) separated by a blotter impregnated with vinegar; the resulting voltage was in the range of 1 V. But these systems can be stacked to associate these generators in series and so obtain higher voltages. These stacks of metal plates were then termed as "piles". Of course, such a system will not deliver its power endlessly but eventually come to a point where the chemical concentrations equilibrate and the power vanishes. From that initial time of innovation, the discovery has led to a prodigious variety of systems adapted to the application involved.

Then, such piles are consumable products; but that does not prevent a very large usage of the piles. Nevertheless rechargeable "batteries" are also available which can be recharged when empty. The size of these systems has been adapted to fit with a huge diversity of usages. There can be very small to be included in tiny devices such as clocks or smart-phones, or very large to store huge volumes of electricity such as batteries used in submarines.

Batteries are chemically arranged structures similar to the piles but with the difference that the asymmetry of chemical concentrations can be restored and the power supply renewed. Batteries deliver direct current and are a very common and practical way to store energy even if somewhat limited in capacity. They can be used as well in medical applications (implants) or even, now, to move cars, trucks and, even, some speak of planes. Robotics is a 2.0 expanding technology which relies exclusively on batteries in order to remain autonomous. Even a new gadget has appeared in our streets: the electrical scooter is all the rage at the moment!

4. Radiations and Sun

The Sun is the star at the center of the Solar System. It is a nearly perfect sphere of hot plasma, heated to incandescence by nuclear fusion reactions at its core, radiating energy mainly as visible light, ultraviolet light, and infra-red radiation. It is, by far, the most important source of energy for life on Earth. Its diameter is about 1.39 million km (864,000 miles), or 109 times that of Earth. Its mass is about 330,000 times that of Earth; it accounts for about 99.86% of the total mass of the Solar System.

It was formed approximately 4.6 billion years ago from the gravitational collapse of matter within a region of a large molecular cloud. The Sun's core fuses about 600 million tons of hydrogen into helium every second, converting 4 million tons of matter into energy every second as a result. This energy, which can take between 10,000 and 170,000 years to escape the core, is the source of the Sun's light and heat. The energy of this sunlight supports almost all life on Earth by photosynthesis, and drives Earth's climate and weather.

Neutrinos are also released by the fusion reactions in the core of the Sun, but, unlike photons, they rarely interact with matter, so almost all are able to escape the Sun immediately and may cross the Earth without interacting with it.

Concerning Energy, a comment must be made about its implication in quantum theory where electrons are represented not as particles with mass but rather as wave functions. This is depicted by the famous Schrödinger equation. The story of why is too long to tell but, trust me, it works! There is the paradigm of uncertainty. Nothing is certain but only probable.

What remains certain is the conversion of the sun's photons into active electrons by the means of photovoltaic conversion when interacting with matter. The use of ground-mounted photo-voltaic (PV) solar cells started in the United States around 1978. At that time, the technology lead the United States had gained from providing power supply to spacecrafts, and consequently the governance of key knowledge, had spurred the first prototypes aiming to develop PV cells as a future electricity source also on the earth. However, such converters required very large collection surfaces.

Various semiconductor materials may be used to convert the sun's photons into electricity, from silicon to zinc oxide with various spectral responses (Figure 2.6) extending from the ultraviolet (UV) to the near-infrared as shown in the plot below. Silicon is largely preferred as it can be mass-produced easily and gives a good yield of conversion.

Such generators are adapted to the intended diversified goals: space applications, solar panels, domestic supply on rooftop panels, large energy farms or autonomous beacons as well. Other applications use concentration mirrors to make the capture more efficient.

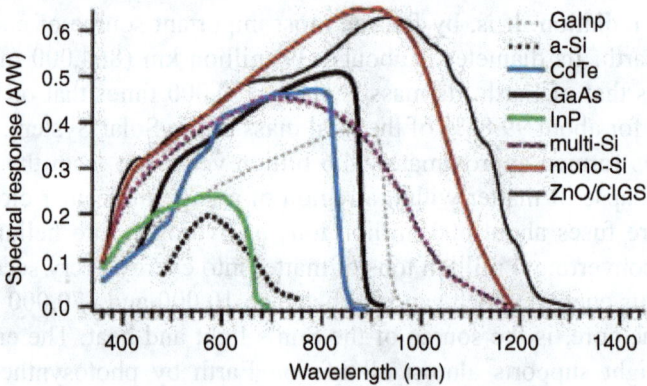

Figure 2.6. Graph showing the comparison of the spectral response against the wavelength for various semiconductors.

Obviously this "renewable" energy is not constant but linked to daylight; so it is worth thinking of a buffer system (batteries). For instance some systems are very comfortable with applications that accept a stop-and-go procedure such as a desalination plant or a water pumping station.

This collected energy contributes to sustainable development but in no way might compete in volume with nuclear plants. That, however, does not preclude an exponential rise of the implementation of such generators.

5. Nuclear

The nuclear paradigm emerged in 1945 with the first atomic bomb on Hiroshima. Since that time it has, hopefully, migrated toward more peaceful horizons without forgetting its original vocation. So what's the matter?

As the theory goes, when matter began to be created after the Big Bang explosion there was a huge convergence of energy which condensed into a multitude of atoms. Some were small, others very large. They all were made of a nucleus and a cortege of electrons bound by two types of forces.

The first one is a long-range electrostatic attraction or repulsion (Coulomb forces) depending on the polarity of the charges involved

$f = qq'/d^2$. The others are very short-range strong forces and are located inside the nucleus in order to keep the pack tightly bound; they are commonly known as "strong interaction" and act on both neutrons or protons; this helps in keeping all these populations together. Any change in the nucleus's constitution involves energy being emitted or absorbed due to these forces, whether the nucleus is small or large.

These atomic elements are usually classified in the famous Periodic Table of Elements by Dmitri Mendeleev (Figure 2.7) as reported here below. The lightest are in the first line, the heaviest in the last one. There cannot be stable atoms at heavier masses because of the repulsion of the protons and the induced instability of the structure. Some nuclei such as ^{235}U or ^{239}Pu (and also thorium which is often forgotten) are able to enter spontaneously into a radioactive decay, emitting a neutron of high energy (19,300 km/s). This neutron, as it is, is unable to stimulate the same process in a neighbor nucleus. To generate a maintained reaction it is required to slow it down to 3.1 km/s by managing several elastic shocks with neutral atoms surrounding (such as water). Then the reaction can amplify, thus giving birth to a nuclear chain reaction[8] in a nuclear reactor or an atomic pile which is accompanied by a heat release.

In a fission reactor, this heat is transmitted to a heat transfer fluid (pressured water or liquid sodium) which can drive a turbine or something similar. Energy is transmitted step-by-step (Figures 2.8 and 2.9) to the place where it can be properly used. This is the classical scheme for all reactors in use. The power output often reaches into the megawatt range, hence nuclear fission is considered a powerful source of available energy. A scheme is depicted here below. Hot water is evacuated and led to a cooling tower giving rise to a vapor plume.

In 2019, some 454 power reactors and 226 research reactors were in operation around the world to help slake our thirst for energy. Some worry about possible accidents of hazardous accidents, such as what happened in Fukushima for instance. Whatever the means involved, manipulating high amounts of energy always remains potentially dangerous whether nuclear or not. These reactors are undeniably dangerous but we need them, and we must continue looking for still greater resources.

Contrary to what is often said by ecologists these reactors do not produce any pollution. The clouds above the cooling towers are only water

---

[8] This is a very schematic explanation.

Figure 2.7.   Mendeleev's periodic table.

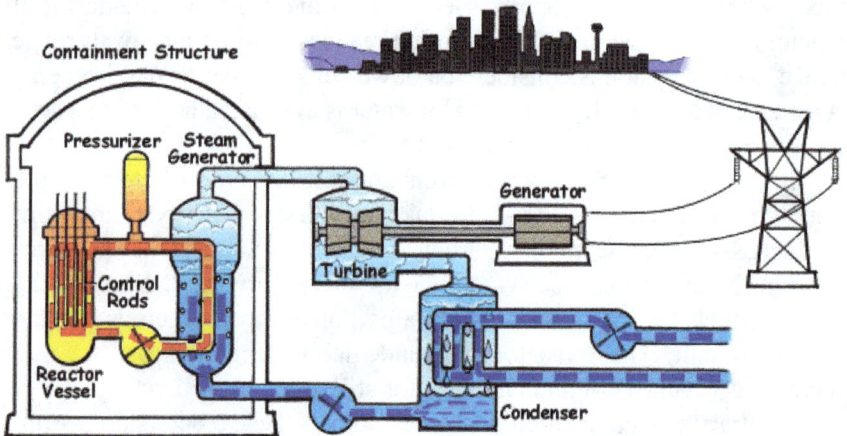

Figure 2.8.   Schematic view of a pressurized water reactor (PWR).

Figure 2.9. Cooling towers and their water vapor clouds.

vapor, unlike coal plants which emit highly polluting smoke. The only pollution they produce comes from radioactive waste which must be carefully buried to properly contain its radioactivity.

As time goes by, it has become necessary to search for a new source of energy to keep up with an increasing demand. Great hopes were focused on another kind of nuclear reaction that involves larger indicial energies: the fusion reaction. As shown in Figure 2.10, the energy per nucleon involved in a fusion reaction with light elements is noticeably larger than that involved in a fission reaction. So this makes for a very promising source of energy.

We do know that this energy can be triggered in a very devastating way: the hydrogen bombs. In this case the initial energy is provided by a "match" consisting in a "classical" uranium fission bomb and the energy is definitively lost.

However, mastering this energy of fusion in a well-controlled way is a quite different issue. Solutions are still years away.

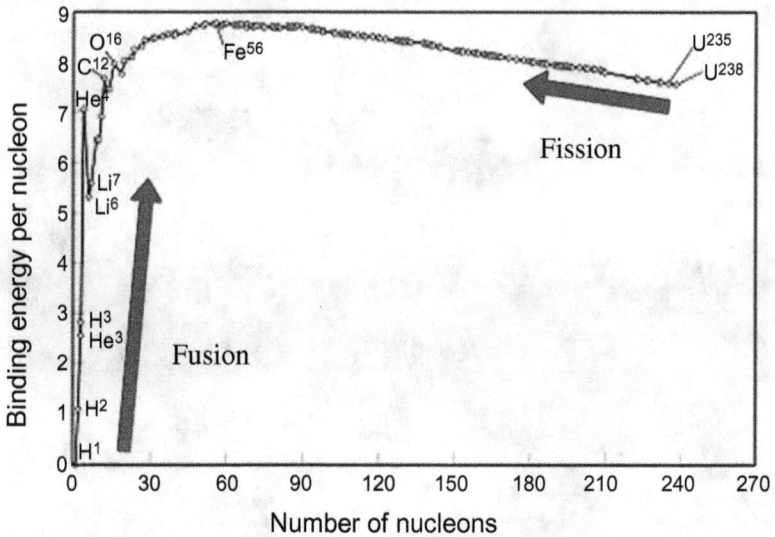

Figure 2.10.   Energy available per nucleon.

A world competition is ongoing with different levels of success. As Maria Zuber[9] said: "Fusion, in a lot of ways, is the ultimate clean energy source. The amount of power that is available is really a game-changer, it's limitless. We just have to figure out how to utilize it". Europe manages the collective project ITER[10] which is to be up and running soon.

Currently, besides Europe and the US, many countries have independently developed experimental set-ups already under test at various levels of sophistication: China (3: HL-2M, EAST, and 3-TEXT), Russia, South Korea, India, and Japan.

To start a reaction between light elements, it is necessary to make them collide with great energy, then they have to be strongly accelerated in a toroidal magnet (Figure 2.11) we call (in Russian) "Tokamak". Theoretically, a fusion reactor should be able to produce an endless amount of energy without greenhouse gases and nearly without any waste.

---

[9] MIT-designed project achieves major advance toward fusion energy. *MIT News.* Massachusetts Institute of Technology. (2021). Retrieved December 9, 2022 from https://news.mit.edu/2021/MIT-CFS-major-advance-toward-fusion-energy-0908.

[10] International Thermonuclear Experimental Reactor. A common project involving 35 nations.

THERMONUCLEAR FUSION REACTOR

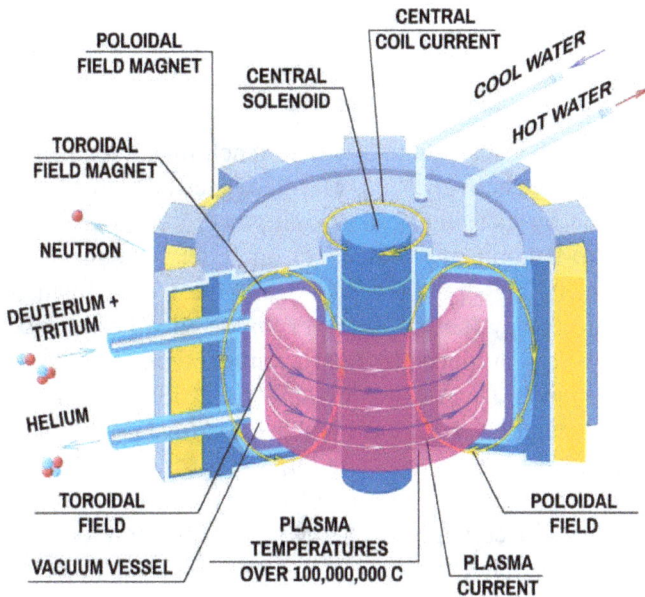

Figure 2.11.   Cutaway view of a thermonuclear fusion reactor.

The basic set-up consists of a toroidal superconducting magnet[11] providing a magnetic confinement (immaterial barrier) for deuterium (^{2}H) mixed with tritium (^{3}H) ionized atoms injected tangentially from an external source. The plasma enters in an accelerated rotation without coming across a material obstacle and so reaches a very high enough temperature in the range of 110–150 million degrees, allowing the start of a fusion reaction.

The main problem, remaining unsolved, is how to extract this energy satisfactorily and put it in a useable shape the day we would be able to outsource more energy than invested.

In China, the reactor "Artificial Sun" reached recently a record temperature[12] of 120 million degrees for 101 seconds and later 160 million

---

[11] High temperature superconducting magnets are now available.

[12] China maintains 'artificial sun' at 120 million Celsius for over 100 seconds, setting new world record. *Global Times*. (2020). Retrieved December 9, 2022 from https://www.globaltimes.cn/page/202105/1224755.shtml.

degrees for 20 s! But the Chinese researchers expect a delay of 30 years to obtain an operational machine.

In 1995 the Fusion Test Reactor (TFTR) of the University of Princeton had already reached a temperature of 510 million degrees (record still valid) whereas the Tore Supra (CEA France) obtained in 2003 stable plasma at 6.5 min.

However there is also another way to reach fusion temperature and it is being investigated actively,[13] that is called "inertial confinement" which can be obtained by focusing several high power pulsed lasers on a small target (a micro-balloon of ^2H and ^3H), in order to manage, in a brief moment,[14] a very high pressure and overcome Coulombian repulsion. From this huge energy (peta-watt lasers) involved we can expect to gain a still higher energy produced by the fusion reaction.[15]

The first positive success was obtained in 2013 and work is in progress in many countries (e.g., the French laser Megajoule, the Japanese Koyo-F, and the German Marvel Fusion).

The day that a positive payout of energy can be extracted from a fusion reaction will certainly be a revolution in our lifetime, one that would provide us abundant electricity at low cost and without pollution. That will be a game-changing event leading the 2.0 age to become a truly new civilization.

6. Space

The 20th century has given us plenty of unexpected new fields, among which space has been prominent.

The chemical energy of propellants is largely used to bring rockets outside the atmosphere and bring satellites into orbit around the planet or even much farther. We are now in the space paradigm and energy is the key word there. If the Moon is to be colonized we should have to find energy sources other than the Sun to ensure decent sustainability of life there.

The number of applications is limitless from weather forecasts to oil and gas prospecting, not to mention the essential domain of

---

[13] National Ignition Facility (California).
[14] 0.1 trillionth of a second.
[15] Nothing is for free!

communications. Technologies for space missions are being made available to address the burgeoning energy needs of "Spaceship Earth". Today, space has a key role in monitoring greenhouse gas emissions and its climate impact.

Collecting energy from space and transmitting it wirelessly was first described by Isaac Asimov in his book *The Roving Mind*.[16] Scientists became interested in this idea in the 1970s. After numerous studies, various concepts have been identified that use all sorts of principles for the production, transformation and transmission of energy, but from a financial point of view, it was unrealistic, so the research did not lead to anything up to now. It is a pity because it would have allowed the collection of the sun's energy in a permanent way.

The problem we face with energy (some say) arises from the increasing Earth population, its increasing longevity and the increasing needs one exhibits as a consumer. Would the solution lie in a deadly virus as begins to be the case today or would we have to migrate to another planet in the future? Perhaps insights might be found by looking back to the past history of discoveries.

---

[16] Asimov, I. (1997). *The Roving Mind*. Prometheus.

Chapter 3

Discoveries Follow Each Other in a Precise Order

Energy is everywhere at work or in life, even the more virtual ones; without it, nothing can happen. However, for things to happen requires background knowledge and adequate means, both resulting from the past. This progress requires the diffusion of knowledge of the art form. No discovery emerges from nothing; there is obviously a chain of events that leads to the final innovation. Now we have looked at every aspect of energy that transform our civilizations (Figure 3.1). We have, then, to invent something else to animate the digital 2.0 world which seems to lack inspiration!

The first men were naked; we, now, are in warm clothes; those did not come in a jiffy but resulted in a progression of the specifications in the making. Christopher Columbus would never have discovered America without using the energy of the wind to pull his ship forward.

Here, we present an overview of the successive scientific insights related to Energy discoveries and their consequences on the longevity of the world population. The initial development of modern civilizations began in Mesopotamia (Sumer) where a coded script was invented because of the requirements of the emerging long-distance trade. That was the very beginning of written languages. Nothing happened by chance but followed a need and an opportunity.

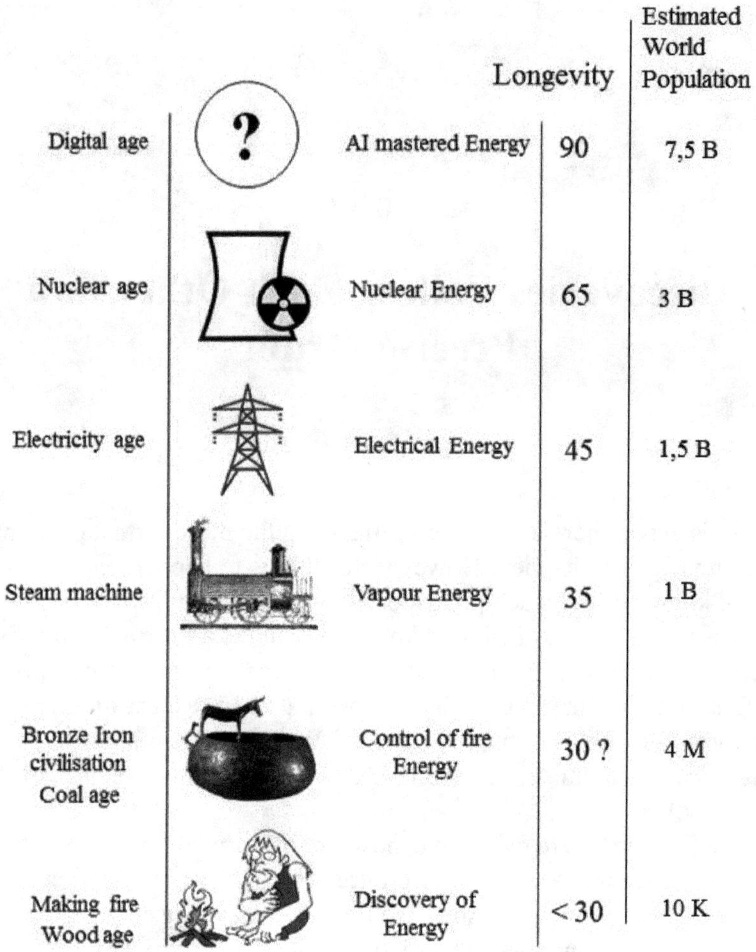

		Longevity	Estimated World Population
Digital age	AI mastered Energy	90	7,5 B
Nuclear age	Nuclear Energy	65	3 B
Electricity age	Electrical Energy	45	1,5 B
Steam machine	Vapour Energy	35	1 B
Bronze Iron civilisation Coal age	Control of fire Energy	30 ?	4 M
Making fire Wood age	Discovery of Energy	< 30	10 K

Figure 3.1.   The main steps of the emerging progress.

1. Some historical examples

To begin, it is a common saying that bringing fire under control was needed before the invention of metallurgy; also building a cart required prior invention of the wheel and so on.

Electric wires were needed to allow Samuel Morse to inaugurate his first electrical signal communication (Figure 3.2) with the famous announcement: "attention the Universe, by kingdoms, right wheels!" in a coded way.

Figure 3.2. The telegraph which can transmit signals via Morse code, such as the distress signal SOS (... --- ...).

That was the very preliminary innovation for a further fantastic development of the digital transmission of information over long distances.

However, the wires were somewhat cumbersome and limited in a one-to-one transmission. Electromagnetic waves then came with Guglielmo Marconi, allowing a broader distribution. Step by step, things improved over time, allowing for more sophisticated applications driven by energy.

This was a major discovery which initiated a number of consequences. Nowadays things have evolved; building the Internet required the prior invention of the transistor and the following diversity of the electronic circuits which manages telecommunications and so many other applications (in the medical domain for instance) which require more and more energy (likely in electrical form). We are definitely deeply dependent on these networks of information transmission which flows nonstop day-in-day-out.

All of that began in 1947, when William Shockley, John Bardeen and Walter Brattain[1] (Figure 3.3) discovered what they called "the transistor effect". They were then working at Stanford University on the electrical properties of some rather exotic materials: semiconductors. At that time it was pure scientific curiosity; little was known about these crystals

---

[1] All three were Nobel Prize winners in 1956.

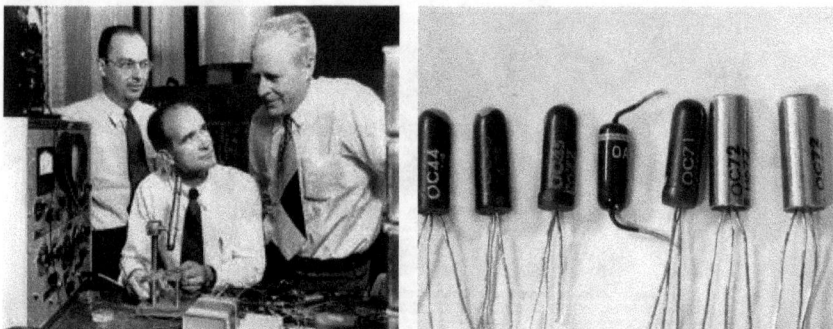

Figure 3.3.   The discoverers and the first transistors.

(germanium essentially) which were able to be conductive as well as resistive, depending on their "doping" with other selected atoms. Their job was to decipher how electrons behaved in these structures of "energy bands".

They discovered that a structure consisting of three zones differently doped could be electrically controlled by three corresponding electrodes, thus making it possible to "modulate" a current following a command voltage. That was exactly a solid-state copy of what a "vacuum triode" did (but required powering by higher energy consumption).

The industry jumped on this serendipity and created the first transistors with the aim of replacing the old vacuum tubes, to minimize the energy consumption and heat. Thus, the first "three-legged" components were born! So tubes were to disappear and transistors later became the field effect transistors that underpinned the development of integrated circuits (ICs), making thus a typical hierarchical implementation of discoveries.

After germanium, used in the first "three legged" transistors, silicon was then preferred because it was more convenient to mass produce them. Ingots were grown as monocrystals in a furnace[2] and reached a more and more extended size, up to, soon, a diameter of 450 mm. Here too the progresses were achieved, step by step, through trial-and-error, interleaved with the joy of success and the sadness of disappointment.

But science is rather eager; silicon has been considered a bit too "slow" for the paradigm of hyper frequency applications and other

---

[2]Following the Czochralski technic.

Figure 3.4.　Typical silicon crystals and corresponding wafers typically 1 mm thick.

materials such as gallium arsenide (GaAs) have emerged (Figure 3.4). As its name tells us, it is a binary compound much more difficult to obtain as a "defect-free" monocrystal, because two different atoms are to be brought together in a convenient order in the crystal, but the knowledge accumulated with silicon helped to take the great leap forward and, gradually, monocrystals, of convenient quality and size, have been produced.

Today all of us may note the fabulous diversity of the applications that followed the discovery of the transistor effect. Our present world would certainly have not existed without this discovery. The following miniaturization of electronic components led to the invention of the current ICs.

The game behaves in a tricky way: as techniques became more effective, the individual components become much smaller, cheaper and energy-saving; but at the same time the opportunity happens to do many more new things than previously attempted… which requires more components with ever-increasing complexity, and the global balance of energy may, then, be questioned. It is known that data centers (comprising thousands of solid-state drives (SSDs)) need to be strongly refreshed and so need to be installed close to a power station to limit energy losses!

2. From the concept of tools to that of systems

Our recent history is fascinating because the motion of "progress" is accelerating woefully whereas, just a century ago, a man's lifetime was hardly enough to perceive "progress". Today our children will see the transitional stages of the evolution unfolding before their eyes. They will discover their steps and the direction (Figure 3.5).

Figure 3.5.   Now hundreds of ICs are implanted on the same "wafer" and then cleaved into individual chips.

For millennia, the notion of "tool" prevailed. The underlying implied idea was to reinforce the abilities of the hand and generate the use of new energies (a hammer for instance). This approach, for the first time, clearly distinguished man from animals and also allowed him to protect himself with choppers, spears, arrows, etc. from predators. The invention of bows and arrows used a stored elastic energy to propel a weapon. Tools were of primary importance to make clothes from animal skins and, later, plant fibers. Each improvement of a tool led to an improvement of the living standards of comfort and efficiency.

Tools became an extension of the hand to improve efficiency, strength, and precision of gesture to obtain an adapted energy. Some animals instinctively use tools to break a nut for instance. However, animals do not make the cause-to-effect relationship between the stroke and the nut that breaks; a glimmer of intelligence is lacking.

Over millennia, the thick brain of the humanoid became open to logical thinking and the management of energy. Human thought became open to the "prediction" and reproducibility of an immediate future (implicitly, the physicist was born). These mutations took time to be adopted but they were irreversible.

Figure 3.6. Steam-powered engine of bygone times.

For a very long time, the intrinsic value (preciousness) of a tool remained very high because of its scarcity and the labour required to produce it. Later, machines and automatisms took over, manual labor became less useful and less efficient.

We entered then the era of machines; particularly in the 18th century when the thermal energy of the water vapor was harnessed, thus giving a new dimension to human activity. The stored energy from compressed steam gave rise to endless opportunities among which we can name the famous locomotive attributed to George Stephenson (1814) (Figure 3.6 and 3.7).

For the first time this locomotive allowed transport of heavy loads over long distances without fatigue; there was no longer a need for horses. The same thing happened with boats which no longer required sails, and thus, became independent of the vagaries of the wind. All of that was due to energy storage, from a long time ago (thanks to the Sun and CO_2), in the prehistoric trees which, over millennia, were transformed into coal. Wind therefore became a useless energy source for a long time and it is funny to see, now, a return to the wind turbines which follow on the old windmills to provide "renewable energy" (when the wind blows!).

Today one prefers petroleum or gas transported over thousands of kilometers along pipelines or with tankers all over the oceans. Some say this is the "lifeblood of civilization". We have definitely left the idea of

Before   Now

Figure 3.7.   Old and new ways to harness the wind.

the candle in favor of oil lamps and later electric bulbs (then successive mutations of the solutions to the same need for light).

A "mechanical slavery" began, resulting from domestication of energy sources, multiplication of forces and removal of the arduousness of work involved. Machines received a form of autonomy when they were equipped with commands to obey to preset orders and programs.

Two centuries were required to enter the new phase of what we call "systems". Machines are connected in a network to perform their complex tasks together. A refinery, a carrier rocket and a car production line all constitute systems. Automation and Artificial Intelligence (AI) dominate the field. It is hard to distinguish what new concept could supersede this state of doing things in the future, and to what direction the evolution of the human behavior could take us. Unquestionably, that will lead to a new style, and it will take a while before we notice that a new age has been born.

Now, technology has given us incomparable means to store, keep, rationalize, allocate or even simulate this providential wealth due to energy disposal. An idea is given here below of the various contributions of the energy sources presently exploited in the world.

Energy and the management of it are everywhere to make a new opportunity follow. Discovering the use of the rudder was required before we were able to sail safely off the shore. In any event, energy consumption is still increasing particularly with the natural resources which have taken millions of years to build up. Some expect (Figure 3.8) that a deep change in our way of using fuels could occur around the year 2050. However, the silver bullet has still not been fired!

World power generation by fuel
TWh

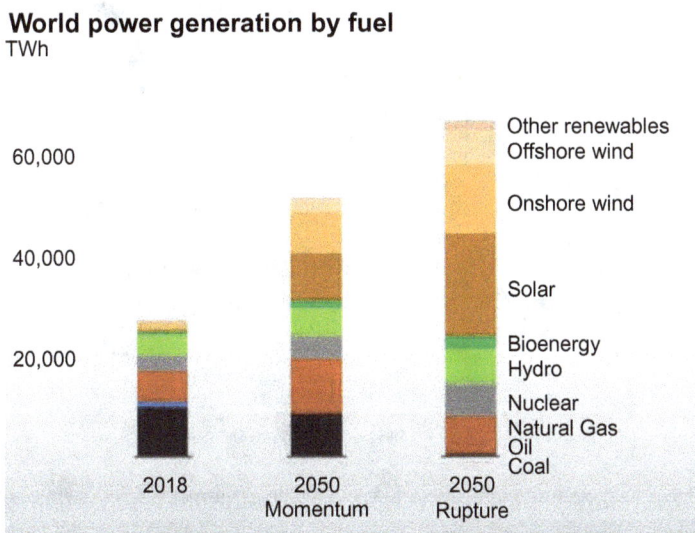

Figure 3.8. Projection of energy usage broken down by source.

3. Counter examples

However, there were also famous counter-examples of discoveries which were not directly exploited by their discoverers but remained "in stasis" before being rediscovered and exploited in another context.

The most famous examples came from China. Fundamental discoveries emerged in antiquity but they did not trigger any direct consequence at the time. It was only several centuries later that other people brought a further follow-up.

Black powder (Figure 3.9) was originally invented (by chance) in China and used to make flares or fireworks. It took a long time for the British inventors to use it for gunpowder with the success we know.

Similarly the magnetic compass was invented during the Han dynasty (206 B.C.–220 A.D.) and was initially used to practice *fengshui* to orient homes correctly with respect to the magnetic field. It only became a navigation tool in the 12th century. Today new versions of the compass use other physical phenomena such as gyroscopic orientation or, more recently, satellite radio guidance.

However, fundamental discoveries such as the wheel and correspondingly the wheelbarrow or the carriage were at the center of every famous architectural innovation such as the Pyramids, but that did not stop the

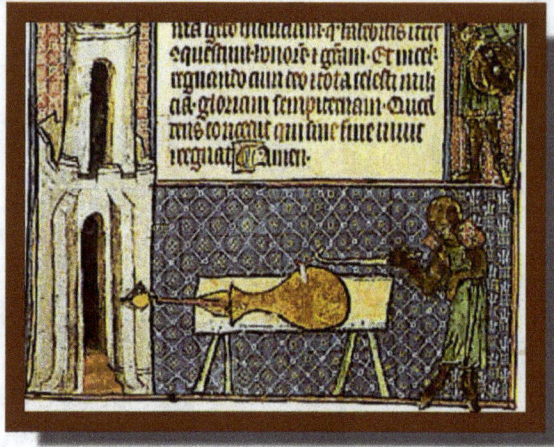

Figure 3.9. The first attempt to use black powder as propellant for a crossbow bolt: Oxford 1326.

Inca and the Maya from constructing similar buildings without the help of any wagons!

Our dependence on progress basically followed the successive discoveries that lead us to new opportunities. Reflection has taken over from physical effort, and technology has followed. The present living environment of the *Homo modernicus* no longer has anything to do with the situation of what it was a mere half century ago.

The last step in the domestication of inert matter is underway: matter is going to become "thinking". Robots are able to make decisions in a way more rational and reliable than humans. Ancestral taboos are over: no more night-time terrors with the electric light, no more ignorance with the large distribution of information. Nobody is now supposed to be ignorant of anything, for the Internet is here. The counterpart arises and man has begun to accept the reality that the sky [heaven?] is empty.

Knowledge is now available to all and no longer restricted to the privileged few. Energy is spread all over. We do not know what the impact of all of that would be on the social equilibrium of mankind now that energy is obtainable for all.

4. The role of information and communication

Information about every discovery or serendipity requires dissemination across a wider scale so that others can use it in every other way possible.

Word of mouth was the only way in the old days but writing soon appeared, then printed books, and encyclopedias to supplement memory and a long-range communication; now the Internet has taken over as a universal means of information. The old days of recording knowledge in the form of printed dictionaries or encyclopedias are over.

Progress is made by riding on the knowledge available; then a network of communication provides the channel to propagate the energy from point to point. However, similar to Energy, Information is a virtual concept which incorporates at the same time any technical knowhow (if any). Information, as energy, is a consumable product which vanishes as soon as it is received, which makes it different from knowledge which results in a quantum of information carefully stored. Our modern life makes us addicted to information; so information is a permanent consumer of energy. This occurs to the extent that the level of civilization in a society could be measured through its volume of exchanged information. It seems to be a vital fluid essential for society.

Just like energy, information can be measured even if it is virtual. This is calculated through the concept of probabilities and the following unit is the "bit" (binary unit). The means of transmission of information quickly evolved from Morse digital code to wireless transmission of the voice (Fessenden) and then television.

Then the "great sorcerer", Claude Shannon (Figure 3.10),[3] came to create mathematical order in the content of word information. These codes he invented helped to initiate trans-oceanic cable links, and allowed effective communication with remote satellites to be established. Voyager has left the solar System and is still active even if it has just 300 watts available in its radio emitter to reply to Earth. This is made possible by the magic of Shannon's codes.

Even robots are now able to exchange information without human intervention through their own network robots on Earth. Now our accumulated knowledge doubles every three days and that requires a wealth of energy for the "data crunchers" to keep the memory of that unlimited material.

For discoveries to happen, they require the right prerequisite and the right opportunity to coincide at the right time, and driven by the right energy. Along this path they need a constant communication of information. Human intuition is the basis of the idea and, to date, AI has not yet

---

[3] In spite of his fabulous contribution, he has never been honored with a Nobel Prize.

Figure 3.10.   Claude Shannon.

been able to evolve into that paradigm located in the deep subconscious of the brain.

The implementation of electricity has been at the heart of a myriad of discoveries previously inaccessible, in a large variety of domains from home energy distribution to machines, cars, heating and other endless applications. It was a major and vital vehicle to distribute energy in large quantities in our march toward the 2.0 world.

The part of electrical energy in the global balance has risen steadily along the years as shown in the Figure 3.11.

There is no natural deposit of electricity; it is obtained by transforming another primary energy such as wind (always wind!), waterfalls, coal (again), gas or petroleum, unless we fall back on nuclear power. Electricity, hopefully, can be stored in batteries even if doing so is costly, temporary and limited. Some batteries are single-use pieces of equipment while other can be recharged. Electricity has become a vital form of energy in our modern world, considering the ever-increasing number of applications and users that followed.

So, electricity comes in two different forms: either as direct current from a dynamo, a battery, a rectifier or as alternating current from a transformer, a generator, or a modulator. Depending on the situation, alternating current is more likely intended for remote transmission of energy

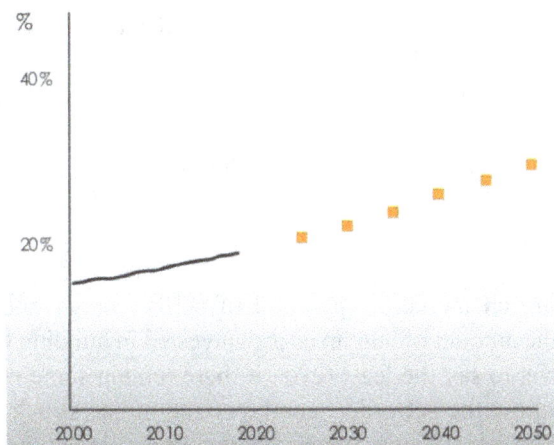

Figure 3.11. Share of electricity in total final consumption.

whereas direct current remains limited to those in closer proximity. Electricity is certainly the most polyvalent form of energy and can be used in unique specific situations such as soldering, ignition in spark plugs as well as light sources, heaters and so on.

Today, electrical cars play a large part in the car market and consequently more and more batteries are needed, and thus their manufacture is required in ever larger volumes. The manufacturing of energy storage is costly and consumes large quantities of expensive "rare earth" materials (lithium, cobalt, vanadium essentially) which, as their name suggests, are rare and easily finite. The main flaw of batteries lies in their limited charge capacity. Also, their excessive weight could be considered a consequent drawback which prevents their use in plane propulsion, for instance. It is also not to be forgotten that, when active, a battery releases hydrogen and that can be dangerous in a closed environment.

At the very beginning there was obviously no electricity, the only requirement was eating. Afterwards other necessities emerged: building a home, covering ourselves, etc. So, a pyramid of needs has emerged transiting from the state of unnecessary to that of essential. That is to say that the personal computer (PC) which appeared quite superfluous some 25 years ago has become a daily necessity today. This new need has appeared and will now never vanish. The same thing happened with the mobile phone which now is universally adopted and has taken its place in our civilization at an individual level.

All these technological opportunities that were before reserved to the privileged few are now finding their ways into our life and becoming essentials that we could not function without, just like food and energy. Nobody can now pretend to be ignorant of anything. Does everybody can assume the risk of having to know?

The invention of the computer is a masterpiece of human achievement, requiring an incredible pyramid of skills, energy and culture. In comparison, the amount of human energy invested in building the pyramids is but primitive, to say the least (even if there remain some mysteries)!

Things follow the set order; computerization has struck the production industry the same way as mechanization underpinned the agricultural industry during the 20th century. It looks like an explosive growth (some would say "exponential", even if not at all) and the pace of Moore's law[4] does not slow, resulting in unforeseen changes. The creation of a new system requires assembling the corresponding knowledge which can lead to the technological solution, with the "simulator" helping to visualize the result.

All are confident in the apparatus that the computer has come up with. A striking illustration of that can be seen in the first flight of the "jumbo" A380: when the test pilot landed after a satisfying flight, he said "That was OK, I had no surprises, the plane behaves exactly like the simulator!"

5. The adverse impact of excessive deployment of energy

The modern society we live in is only half a century old; we lack the vision to see and project forward; but we will no longer have this distance because of the speed of current evolution.

However, the excessive global development of the implementation of massive energy deployment needs, at present leads to disastrous and unwanted side-effects. Accidents involving oil tankers at sea produce

---

[4]The number of transistors in a dense integrated circuit is said to double about every two years.

large slicks of pollution, which are difficult to intercept and neutralize. Induced global warming is known to contribute to uncontrollable summer forest fires, in Amazonia as well as California.

Nevertheless the "progress" leads to an increasing longevity and thus an increasing global population of consumers who require ever more energy. Every unit of energy consumed leaves behind unconverted waste (physical, chemical or organic pollutant products). Now people worry about air pollution induced by human activity, excessive CO_2 emissions, plastic waste at the bottom of the sea, supposed induced climate change, the resulting depletion of natural resources and so on.

Today, for societies to be efficient and productive, they require more and more workers and activity or risk the penalty of falling behind in the global competition. Accordingly, they are condemned to grow continuously using more and more energy.

The problem of excessive CO_2 is rather tricky because this gas by itself is in no way is a poison, but is rather a necessary raw resource for all vegetation on Earth (Figure 3.12). It is at the center of the chlorophyll generated by plants and especially trees; then the obvious remedy for this trouble is very simple: plant trees instead of clearing or burning forests. On a global scale, large forests as we can find in Amazonia, California or Asia are the best trap for excessive CO_2.

The massive recovery of this abundant CO_2 has become a technological research topic: algae and plankton feed on this gas and, after

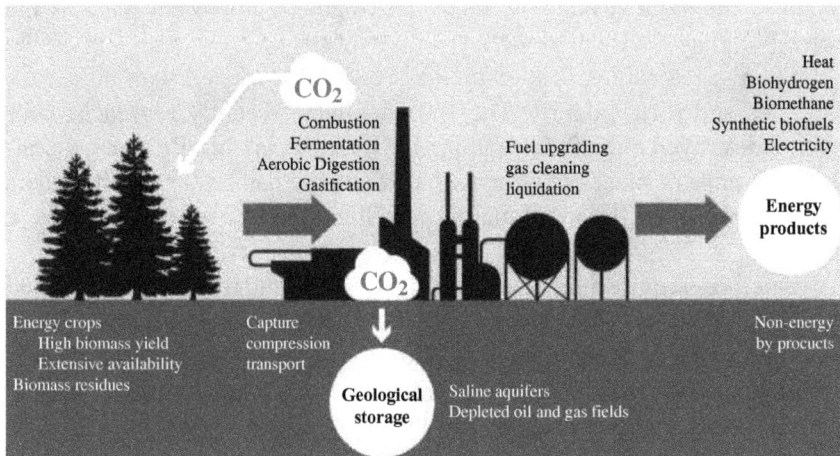

Figure 3.12.   Possible recycling solutions for atmospheric CO_2.

maturation, can be converted into bio-fuel. The circle is completed! The only problem remains the implementation or realizations of conversion plants at a wide enough scale. Another solution would be to inject CO_2 into an aqueous acid solution and filter it with a special membrane to turn the solution more acidic, thus generating both electricity and hydrogen.

However, we must see things in perspective; the deployment of energy due to human activities, even if worrisome for the time being. Natural, uncontrollable manifestation of wasted energy far surpasses the human contribution to the balance: tides, winds, earthquakes, volcanos, storms and so on. It is a pity to be unable to capture this huge amount of lost energy and find an efficient use for it.

The present abundance of means induced by the omnipresent available energy (even if still insufficient) in a conveniently deliverable form has made us fully dependent on the modern way of life. The networks of our societies and their multiple interconnections make them fully secure but, and at the same time, a fault-line in the event of an unexpected danger such as an energy failure.

Robots are now becoming ever more important in the civilization 2.0 that will eventually take over hitherto-sufficient human employment. It is not just big factories that use them on a large scale (to make cars, for instance), but also mass-market retailers (such as Amazon) whose robots are fully independent, their operation and functioning driven by AI. That artificial tireless organization has fully taken the place of the human operators until now; it can work days and night without any vacation or union to initiate a strike! The only inescapable requirement remains the corresponding electrical power!

Even space, which is so wide, has become congested with numerous satellites and rocket debris. The Russians were recently reproached for having destroyed one of their satellites with a special missile thus generating thousands of pieces of debris which could be dangerous for the others and especially the International Space Station which is inhabited by a crew.

Satellites require a lot of energy (provided by the rockets) to be placed in a convenient orbit around the Earth but, when there, their movement is only inertial (a permanent fall) and do not consume any energy for their propulsion. The only energy they require corresponds to their internal functions and communications, and this is mainly provided through solar cells and batteries.

As a satellite (but little bigger!) the Moon obeys the same laws of inertia but its important mass acts upon the Earth and attracts the mobile waters of the oceans, giving rise to the tides.

Would a new human civilization be developed on the Moon, Mars or some other planet? They certainly would require generating, on the spot, their own energy which, at the moment could not be anything other than converted solar radiation (or nuclear power units to be brought there).

Among the various sources of natural energy, coal has long been the most used and that is not going to change soon. New sources of energy stack up over time without removing the previous ones, especially coal which always remains so useful. However, coal and also gas and petroleum ask for a successor which, at the moment, has not clearly appeared in spite of the many "renewable" energy ideas suggested or proposed (it remains an illusion to believe they could be obtained for free!).

The only exception to that rule of the severe energy quest lies in Iceland (Figure 3.13) which, contrary to its name, is a potentially hot country; a country blessed by the gods! No need for a nuclear reactor or wind turbines, the hot water comes from the ground in abundance; night

Figure 3.13.   Hot water from a geyser in Iceland.

and day without interruption water from the dams and hot water from the ground provide megawatts in abundance and freedom from taxes!

It was a unique occasion to motivate some energy-hungry industries (such as aluminum production) to settle there. That is the best of the best!

In our 21st century, a new civilization is yet to be invented to the size of what is needed. For the first time in History a civilization has been born (2.0) without the precondition of a new type of energy source.[5] A super computer requires a power of 40 MW whereas a brain just needs 0.1 W and that is already plenty. The Anthropocene Age is now vanishing; will a Robotocene Age succeed it?

6. Human energies

There are also energies which can hardly be measured in KWh: these which belong to human behavior (Figure 3.14). The same word is used by analogy. Humans are somewhat impressed by the motivation and the desire for self-improvement depending on the characters of people and their courage to undertake tasks.[6] Mental energy is required to carry out a specific hard task and overcome present difficulties in spite of failures and barriers.

Figure 3.14.   A symbolic view of the human energy.

---

[5] Rhodes, R. (2018). *Energy, A Human History.* Simon & Schuster Paperback.
[6] So is the case of writing this book!!!

Food is our main source of physical energy. In this field, there is the modern science of nutrition with its calories, vitamins, carbohydrates and etc. The other source of energy in our day-to-day life, of which we are not fully conscious, is interactions with people which involve exchanges of vital energy.

As what a battery does, human energy (mental as well as physical) may run out, and then comes a breakdown. Our muscles as well as our brain can be exhausted if the bio-energy (measured in KWh) included in the food is consumed. For we are living in a dynamic world that is constantly changing and evolving; evolution and progress make up one of the eternal and universal laws of life. Of course, bio-energy can be stimulated by tea or coffee, drugs, stimulants... or a real "lift me up!"

It should not be forgotten that discoveries, originally, proceed from human activity. They obey a logical order of knowledge and thinking. These shape the tools that are used to execute that human activity, with the tools changing with the available knowledge.

Chapter 4

The Historical Implementation of Energy with Tools and Machines

From the earliest days to the 2.0 world, the physical methods and techniques that humanity developed have been a direct consequence of an adapted use of available energy, and also the common knowledge of the physical world existing at that point in time.

1. The evolution of the means directly drives the results

It is always interesting to study the first ages because they open our eyes to how modern "civilizations" evolve over time.

The key starting discovery was related to a hard, siliceous and sedimentary stone that could be found almost everywhere on the ground: flint. That stone was selected for the specific property of being able to be cleft by striking it with another stone, thus developing very sharp edges. That technique gave rise to the first man-made cutting tool (Figure 4.1).

Human skill soon brought the design of a two-sided shape specifically adapted to the human hand, thus making this stone a genuinely efficient cutting tool. Otherwise, a scraper useful on one side could also be crafted to scrap the fat from animal skins. These tools were made by hand and profitably used throughout the period of utter deprivation that marked the Prehistoric Age.

As such things as arrow heads and buttons for coats were invented, their production became, then, for the scale of the time, a real industry.

Figure 4.1. A flint stone on top, and the corresponding carved one-sided scraper at the bottom-left, and a two-sided knife at the bottom-right.

With such arrows or spears, it became possible to kill prey remotely with more efficiency than the previous fire hardened tips.

Another particular property of flint stones is that when struck with a metal piece, they produce bright hot sparks which enable the lighting of a fire. This was especially useful in those remote times when lighting a fire was a daily problem. In 1991, the frozen body of a prehistoric hunter was found in the Alps at an altitude of 3,210 m; this iceman has been named Ötzi and his body was well-preserved in extremely good condition due to the low temperatures maintained all over the millennia on this mountain. His bag contained just such a lighter.

All along our history, such lighters were widely used and even adapted to the first firearms such as muskets. Such small quantities of energies involved were enough to light the powder and propel the bullet.

The symbolic power of the flint stone (as a talisman) is also known to favor mental energy, courage, strength and fighting spirit. Such a stone dispels fear and re-lights the inner mental flame.

Figure 4.2. First train in La Ciotat filmed by L Lumière.

Later, the discovery of metals likely happened by chance in the residues of a coke furnace where bits of fused metals were found. Then they were hammered into shape or, later, melted and cast. Things developed step by step, either by chance or design. Metallurgy was born in Mesopotamia, giving us a profusion of ever-increasing sophistication in the development of tools, arms or various pieces which can be possibly melted again, afterwards, to create perhaps another object.

Metals appeared in a given order following the ease of manufacturing: copper first, later bronze (sometime in 2000 B.C.) for a long period, and finally iron which was the more difficult to produce (because of the higher temperatures required) but most appealing for its wider usability. Meanwhile, gold was also discovered, producing plenty of jewelry. The age of metallurgy corresponds to a larger prioritization in societies following the emergence of trade and also the subsequently the wars that followed. The key word remains "evolution" and that has not changed up to the present.

Machines entered our daily life, even if it sometimes happened that these new inventions undermined and threatened the livelihood of ordinary people as the case in France when the first train arrived in the station of La Ciotat and was filmed by Louis Lumière himself (Figure 4.2).

2. War machines

It is especially distressing to recall that, along with hunting, one of the most notable human activities from the earliest times has been to make war with all and sundry to seize goods or assets,[1] and that has not changed even today! To guard against aggression, humanity invented strong walls and castles, and, consequently, they also invented machines to break these walls because their hands could not develop enough energy!

Catapults or trebuchets were ballistic devices invented around the 5th century B.C. and designed to throw heavy stones against the walls or any obstacle to open a breach. The energy comes from elastic pieces of wood or from counterweights installed on a pendulum that swings the charge. They were later widely used by Greeks and Romans and during the Middle Ages.

When gunpowder became available, guns and cannons were used to do the same work with a greater efficiency and reliability. Special vehicles, tanks, boats, planes and even submarines are the modern tools for the same purpose: to defeat the enemy

Today, war machines utilize the most advanced technologies of propulsion, weapons, communication and even artificial intelligence to manage the strategies of fight. Space has taken its place for a better observation and guidance of the missiles.

3. Modern tools

Electronic, software, computers behind the development of sophisticated robots that outperform humans in surgical procedures for brain, eyes, heart to undertake the precision and coordination controlled by artificial intelligence to perform their operations (under human supervision, for the moment).

Optics was, for a long time, a collection of empirical knowledge, not a science. One knew that looking through a glass bottle filled with clear water gave strange images as well as glass bubbles, but there was yet no explanation of why. Refraction was known from the time of Ptolemy, but

---

[1] Henry Fosdick (American pastor) said: "the tragedy of war is that it uses man's best to do man's worst."

the laws of refraction themselves not until Snell and Descartes, and the wave description of the phenomenon not until Fresnel and Huygens. No mention of the photon which carries a quantified energy (after Planck's and Einstein's $E = h\upsilon$).

An important contribution to this development was the result of the invention of microscopes. Tremendous gains were obtained following the initial invention of the optical microscopic vision by Antonie van Leeuwenhoek who discovered the microscopic world (Figure 4.3), and more especially the spermatozoids (1677) as the origin of life (then he demonstrated that God is (almost) for nothing!).

When the characteristics of lenses were better understood, it became clear that the magnification of images could be controlled with the use of combination of lenses with different focal lengths. Progress in the resolution of these instruments reached the limits of the diffraction in the range of a half micron for visible light. Seeing microbes was now easy, but optical microscopes were unable to display images of viruses. More energetic photons close to the ultra-violet range are required to further improve the performances, but the limits of the optical microscope were reached and different technology (more energetic technology) was required for further enhancement to explore the microscopic world; the photon had to make room for electrons and a completely different technology: electron microscopy.

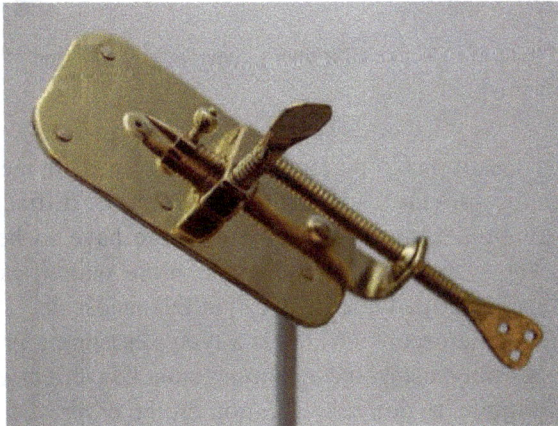

Figure 4.3.   Yes! This is the first microscope! Incredible!

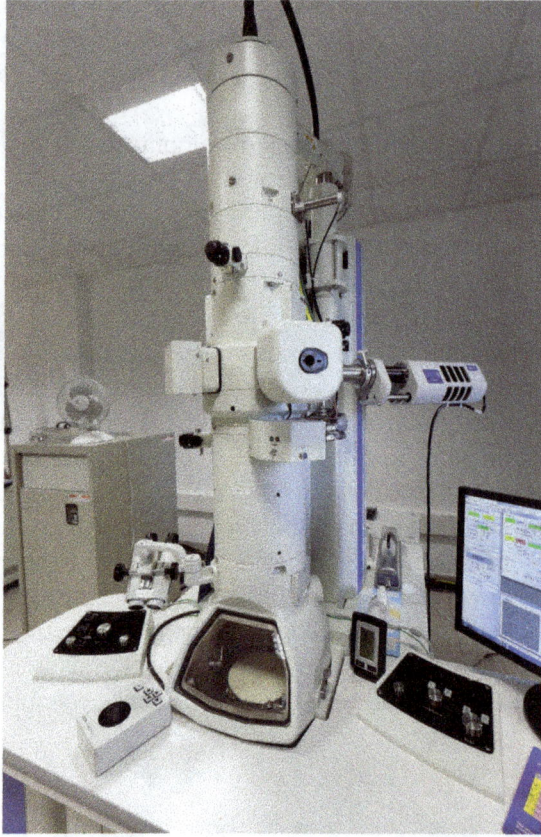

Figure 4.4. A modern electronic scanning microscope, far from the initial van Leeuwenhoek handwork!

Of course, electronics, computer technology and image processing are widely used to provide a detailed image from the diffracted electron beam and resolutions close to the atomic range have to be developed (Figure 4.4). Yet even that is not enough; more magnification is still needed to realize integrated circuits (ICs) at the nanoscale.

Atomic force microscopy (AFM) is a type of scanning probe micros-copy (SPM), possibly using the quantum tunneling effect between the probe and the sample, to deliver resolution on the order of fractions of a nanometer, with magnification 1,000 times more powerful than the optical diffraction limit (Figure 4.5).

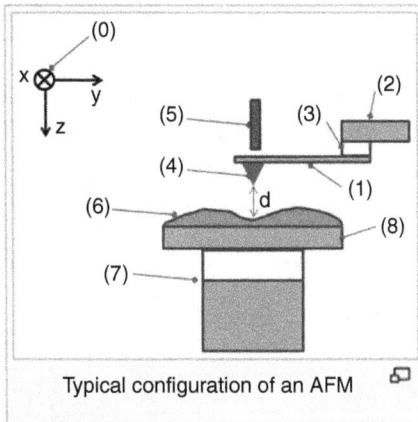

Figure 4.5. Atomic force microscopy.

Note: **(1)**: Cantilever, **(2)**: Support for cantilever, **(3)**: Piezoelectric element (to oscillate cantilever at its eigen frequency), **(4)**: Tip (Fixed to open end of a cantilever, acts as the probe), **(5)**: Detector of deflection and motion of the cantilever, **(6)**: Sample to be measured by AFM, **(7)**: xyz drive, (moves sample (6) and stage (8) in *x*, *y*, and *z* direction with respect to a tip apex (4)), and **(8)**: Stage.

The information is gathered by "feeling" or "touching" the surface with a mechanical nanoprobe. Through manipulation, the forces between tip and sample can also be used to change the properties of the sample in a controlled way (for instance displacing an atom). Examples of this include atomic manipulation, scanning probe lithography and local stimulation of cells.

By so doing, micro-tools called micro-electro-mechanical systems (MEMSs) are an extrapolation of mechanical machines at a micro scale (Figure 4.6) using transistor technology methods: micro-actuators, micromotors and so on. The energy involved is rather minimal.

In a different domain, a laboratory fancy for photon manipulation has led to a wealth of tools of various size and power: these spring from the LASER (or in its full form, Light Amplification by Stimulated Emission of Radiation) invented by Theodore Maiman in 1960 (Figure 4.7). Photons are excited in a cavity where they bounce back and forth until they amplify the beam in a coherent way; this can be performed in a solid crystal (one of the first crystals used was pink ruby), a gas (CO_2) or a solid state diode. Lasers can be operated in a continuous or pulsed mode

An electrostatic "comb-drive" is a common MEMS actuator
used in gyroscopes, microengines, resonators, and other applications.

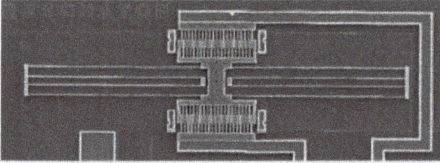

This gear chain converts rotational motion (top left) to
linear motion, thereby driving a linear rack (lower right).

Figure 4.6.   Example of a MEMS actuator.

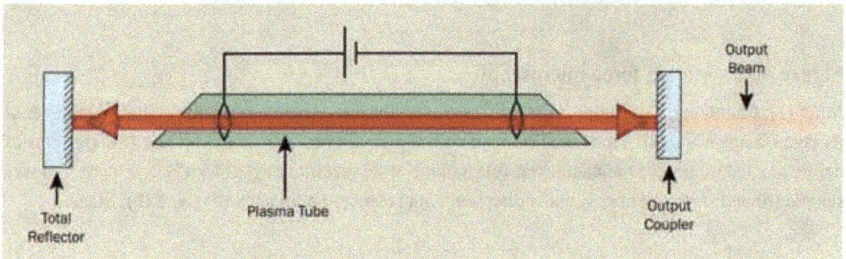

Figure 4.7.   A schematic example of a gas laser.

in a very short and energetic way, in various wavelengths. This makes
this invention a prodigious tool!

This paved the way for a wide spectrum of laser applications from
telemetry, communications, welding, medical, printing, to integration in
solid state devices and many others. This is a typical energy conversion
cascade.

4. Implementation of modes of transport

From the beginning, one of the most important efforts to change societies resulted in the implementation of transportation means (Figure 4.8).
Initially, the range of displacement allowed to man was limited simply
by his legs (using a moderate biological energy). A first improvement
was to use the legs of more powerful animals such as horses, oxen or
camels; that was quite a limited solution, but nevertheless a valuable
assistance at the time.

Figure 4.8. The idea of light as weapon in warfare: Archimedes would have been very excited with this tool to repel the Roman boats in Sicily.

The next great leap forward was the development of transport powered by an external energy source such as water steam heated in a closed vessel, but the real improvement came from the fuels ignited in various combustion engines (with pistons or turbines) that we know today.

This has transformed our modes of travel in a rather short time compared to the previous steps which have evolved over centuries. Initially came the invention of boats, later followed by trains which provided the means for long range transportation of heavy loads. Then individual cars have become commonly used by all. The evolution of the production of various models that followed each other has now culminated in a heap of cars in the cities and elsewhere.

Fuel consumption has soared (Figure 4.9), resulting in uncontrolled pollution and CO_2 emission which is now considered alarming as it is added to the unescapable pollution due to industries. Some also add that an exhaustion of fuel resources may also occur sooner than forecasted in spite of the research extended all over the world (and the seas).

A solution to these plagues was proposed recently to replace combustion engines by electrical ones, but that implies some drawbacks which are often underestimated: the cost of batteries (rare earth elements needed), their reliability, their limited life and stored energy, and the

Figure 4.9.   Petroleum exploration abounds to satisfy the world demand.

corresponding reduced autonomy. This also has the disadvantage of requiring huge amounts of electrical KWh which have to be produced.

However, there are some applications of electrical energy which are the only way to cope. Submarines, for instance, are not able to provide sufficient breathing to a combustion motor (except with a snorkel in order to refill the batteries). Also, trains benefit from a suspended electrical line all along the railroad, and so can be directly supplied from an electrical plant.

A final possibility could be the production of hydrogen which could be considered the ideal fuel, producing no pollution at all. However, that is not so simple because H_2 has to be produced by hydrolysis and this requires lots of electrical KWh. On top of that, H_2 is a gas (a dangerous one)[2] which needs to be compressed or liquefied down to very low temperatures in order to be transported, not to mention the complex distribution networks required. Nevertheless, many attempts are being made to adapt H_2 engines to buses, cars and even planes!

One of the most impressive breakthroughs in the history of transportation was the implementation of airplanes. Nothing of man's own doomed

---

[2]H_2 was responsible for the disaster of the giant transatlantic airship Zeppelin Hindenburg; May 1937.

Figure 4.10. The first flight of Eole (Clément Ader) which was later called "avion", powered with a steam engine.

Figure 4.11. Concorde, such a beautiful bird!

flying machines predisposed him to imitate birds (Figure 4.10). The early attempts were rather pathetic with flapping bird-like wings. There was no real progress until the invention of the propeller which transformed a rotating movement into a displacement of air, giving rise to a force of propulsion.[3]

Of course, significant technological progress was achieved during these historical times so as to improve the performance of these propellers, until the idea emerged to confine the propeller to a duct and design a turbine. Thus the jet engine was born.

Today jet-powered planes cross the oceans at 10,000 m above sea level at a speed of more than 800 km/h. On top of that, a famous plane (Concorde — Figure 4.11) was implemented for commercial fights at supersonic speed but was soon abandoned because of its excessive fuel consumption and after a stupid tragic accident.

---

[3] In some way, the counterpart to the blade of a windmill!

5. Energy efficiency implementation

Every implementation of energy is driven by the need to accomplish an objective, but often the energy required ends up larger than the minimum estimation; this implies expending an excess energy which is often fully useless but unavoidable. Most of the time this loss takes the form of heat that needs to be extracted. This is sometimes called "fatal heat". This, for instance, is the case in a car engine as well as in a factory or in a data centre. It may happen that this lost energy could be transferred in a less demanding application especially when the amount of that energy is important.

In any case, reducing such losses always is of primary interest. For instance, we need a convenient home heating, especially during wintertime, therefore, it is important to insulate our houses carefully and thermally to prevent air drafts or stray heat loss.

Much has been achieved in recent years in architectural and building construction to minimize energy losses (Figure 4.12). The additional cost incurred by this requirement will be quickly compensated from the savings in running costs.

In the development of electronics, the replacement of the old vacuum tubes with transistors and ICs represented tremendous energy savings. In the same way, light emitting diode (LED)[4] components successfully

Figure 4.12.　Thermal detection of heat loss can be easily achieved.

---

[4]Light emitting diode.

replaced old filament lamps with an improved emitting yield and nearly no direct heating, and are widely available at a micro-scale level thanks to mass produced. The life of these emitters also was substantially increased. in spite of the research extended all over the world (and the seas).

In the LED, the recombination of electrons and electron holes, when injected in a depleted zone of a semiconductor junction or quantum well, produces light (be it infrared, visible or UV), through a process called "electroluminescence". The wavelength of the light depends on the energy band gap of the semiconductors used. Since these materials have a high index of refraction, design features of the devices such as special optical coatings and die shape are required to efficiently emit light.

Unlike a laser, the light emitted from an LED (Figure 4.13) is neither spectrally coherent nor highly monochromatic. However, its spectrum is sufficiently narrow that it appears to the human eye as a pure (saturated) color. The white emitted color can be obtained either by mixing the emission of three diodes (blue, green, red emitters) or using a phosphor-based material to convert the monochromatic light of a blue or UV diode to a broad-spectrum emission.

Light Emitting Diode

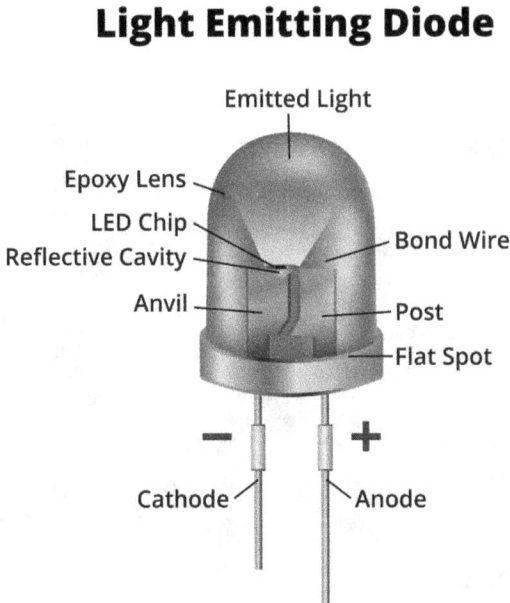

Figure 4.13.   The basics of an LED.

6. Energy and space activity

Access to usable forms of energy is a key strategy when operating in space. Energy is required to propel rockets, to put a satellite into the correct orbit, to power any device or to heat an inhabited area.

Some say space is empty, but that is not true because besides stars, planets and other matter aggregations, space is filled with radiation. Energy in space is a very broad topic. We shall subdivide this study into a number of subsections:

* The immediate neighborhood of the Earth and its connection with human activities.
* The outer space of stars and planets.
* The origin of things (next chapter) and the observed radiations.

7. The neighborhood of the Earth

First of all, the Earth is surrounded by a magnetic field (Figure 4.14) originating in the rotating plasma in the bulk of the planet. North and South poles may be easily located with a simple compass. This magnetic

Figure 4.14.   The Earth's magnetic field.

field is rather efficient in deflecting charged particles coming from outside and, thus, giving rise to the *aurora borealis* in the northern countries.

Satellites can be launched by rocket powered with some form of energy, to put them into the chosen orbit. Thereafter, once in position, they rely on the renewable and accessible energy from solar cells (or, formerly, special radio-isotope reactors). Initially, the idea was to use stations to collect solar energy on a massive scale while permanently illuminated[5] and re-transmit this energy towards a terrestrial station with a carefully focused laser beam, but experimental findings proved unsatisfactory and the idea was eventually abandoned.

Now, near space is densely populated by large numbers of inhabited space stations or lighter satellites (exploring Earth's climate, geography or agriculture) to support the global positioning system. More recently, many "nano" satellites have been positioned (at low cost) in low orbit for short-lived missions.

This region of space in the vicinity of Earth is cold and dark but is also flooded constantly with electromagnetic energy. All stars in the universe produce energy which they radiate into outer space, just like sunlight. Some are accompanied by particles of matter (electrons, positons or even α), or manifest as electro-magnetic radiation (γ), neutrinos or more complex sub-nuclear components.

These energy radiations come from various sources: galactic emissions coming from outside the solar system (nuclear components, solar particles, rays trapped in the earth magnetic fields, solar winds from coronal mass ejections). The Sun is a star which produces its own energy and particles that radiate into in space.

Van Allen Belts were detected first by Explorer 1 in 1958; they are arranged parallel to the magnetic equator and trap incoming radiations to protect the Earth. The lower belts are essentially composed of protons whereas the external one is dominated by electrons.

Virtually everything in space is constantly moving, so there is also kinetic or motion energy in space. All stars or planets are subject to the same laws of gravitation discovered by Newton, driven by their respective masses and speed. Our Moon is the nearest celestial body and unique natural satellite rotating around the Earth. The Moon is the nearest place for humanity to settle and could be considered as a possible relay towards farther destinations.

---

[5] Geostationary orbit is located some 35,000 km from Earth.

Figure 4.15. The reusable spaceship as transport for habitats and cargo for space exploration.

Lunar surface operations are logical destinations in our exploration of the solar system, with the goal of establishing a permanent human presence. Figure 4.15 illustrates the use of a reusable spaceship to provide crewed access to the surface, as well as cargo transport for habitats.

The Moon already shares gravitational energy with the Earth, responsible for the tides generated in the oceans. Its steady motion and phases have long been observed and used to create calendars. This natural satellite is largely illuminated by the Sun but its motion (translation plus rotation) is such that the same face of the Moon always faces the Earth.

8. The outer space of stars and planets

Outer space is an unfathomable mystery that remains to be solved. Its limits might overlap with the extending Big Bang frontier that we are currently trying to explore with the biggest satellite telescope James Webb, which has now reached its final orbit some 1.5 million km from Earth. By then, it might be possible to observe, from a distance, the further

Figure 4.16.   A panel sent out to collect particles in the solar wind and destined to come back.

part of the Universe connected to the earliest times of the Universe to learn about the "Cosmic Microwave Background Radiation"[6] which bathes the Universe at a wavelength of some 4 μm[7] in the infrared (Figure 4.16), coming from every direction in the sky. This energy is explained as the remaining amount of energy originating some 380,000 years after the Big Bang explosion. This fossil radiation is known as a kind of electromagnetic background noise.

Probes have already been sent out to explore this remote space and to map radiation *in situ* and obtain pictures of the stars. Some carried robots which landed on planets like Mars to carry out surface analysis and possibly bring back samples for future analysis.

They carry their own sources of energy (solar cells or even nuclear power sources) to maintain telecommunication with Earth. Special codes have been developed to provide a distant radio-electrical connection.

---

[6] Predicted by Gamow, Alpher, and Hermann.

[7] Corresponding to a temperature of 2.7 K of the Black Body. It was discovered by A. Penzias and R. Wilson in 1965.

Even images are now able to be transmitted over very long distances. A huge variety of instruments of exploration was then sent into space bringing a wealth of information that we could collect and store in data centers.

At present, the farthest space probe mankind has constructed and launched from Earth is Voyager 1, which was announced on December 5, 2011 to have reached the outer edge of the solar system, and entered interstellar space on August 25, 2012. Deep space exploration further than this vessel's mission is not yet possible due to limitations in propulsion technology currently available. Beamed propulsion appears to be the best candidate for deep space exploration currently available, since they are developed based on existing physics and proven technology.

NASA announced the selection of eight American astronauts who will begin training for future deep space missions beyond low Earth orbit for future Mars or asteroid travel.

Celestial mechanics[8] is a branch of astronomy that describes the inertial movement of the massive bodies in outer space. This science elaborated by Kepler, Newton and Lagrange abundantly using the mathematical properties of ellipses was later completed by Einstein who explained the anomalous precession of Mercury's perihelion in his 1916 paper *The Foundation of the General Theory of Relativity.*[9] This paved the way for astronomers to recognize that Newtonian mechanics was not always the most accurate. Binary pulsars have been observed, the first in 1974, whose orbits not only require the use of general relativity for their explanation, but whose evolution proves the existence of gravitational radiation, a discovery that led to the award of 1993 Nobel Physics Prize.

Celestial mechanics in the solar system is ultimately an *n*-body problem, but the special configurations and relative smallness of the perturbations have allowed quite accurate descriptions of motions (valid for limited time periods) with various approximations and procedures without any attempt to solve the complete problem of *n* bodies. Examples are the restricted three-body problem to determine the effect of Jupiter's perturbations of the asteroids and the use of successive approximations of series

---

[8]Fitzpatrick, R. (2012). *An Introduction to Celestial Mechanics 1st Edition*. Cambridge University Press.
[9]Einstein, A. (1916). Die Grundlage der allgemeinen Relativitätstheorie. *Vierte Folge, 354*(7), 769–822. https://doi.org/10.1002/andp.19163540702.

solutions to sequentially add the effects of smaller and smaller perturbations for the motion of the Moon.

Numerical solutions of the exact equations of motion for n bodies can be formulated. Each body is subject to the gravitational attraction of all the others, and it may be subject to other forces as well.

These initial explorations into space demonstrate that however impressive the scales on which our tools operate, these tools are ultimately governed by tried-and-tested physical laws as discovered through painstaking experimentation.

Chapter 5

The Pivotal Role of Knowledge

The notion of energy is purely a virtual concept created to give a basis for a physical evaluation of the world we live in. From there, it becomes possible to accumulate conclusions we gather as a heritage in our memories and call them knowledge as soon as they have been verified. More knowledge means more interconnections and faster progress. Such knowledge is an accumulation of individual thoughts, often erratic and erroneous, lost and later retrieved, but globally verified and tested in reality. Now, advanced technology provides us with un-matched means to store and retrieve, tidy up, streamline, allocate or even simulate this accumulated manna from heaven. The efficiency of this process is a remarkable step forward, far beyond any previous step including the advent of printing.

1. Physical laws deduced from experiences

When man's brain began to develop, the evidence of some elementary physical facts had been appreciated and it was considered that these facts were reproducible. For instance, striking a nut with a stone will certainly break it. From then on, a "law" (coming from Nature) of inductive generalization from concrete objects to concepts was developed: the starting conditions being the same, the result has to be the same. This is the first requirement for reproducibility. Vernor Vinge said: "humans may not be the best characterized as the tool-creating animal, but as the only animal that could figure out how to cognitively reason from observation to conclusion, lay down the foundation for knowledge to develop, consolidate and propagate — beyond themselves into the outside world".

Observation remains the basic origin of the physical laws, and the idea of following the flow of energy is the key track to understanding. Heat has long been a background issue as it was an obvious kind of energy, and entropy was imagined to quantify the atomic disorder in a gas or a liquid, any medium where particles are free to move, under some constraints. Each transformation of an assembly increases its entropy, thus transforming the energy "top range" into a low one. Entropy, then, is the tendency of systems, both in nature and in culture, to run down; so extropy is the human will and ability to use intelligence and tools to improve in developing better systems. Before the term extropy, this was usually described as negentropy or negative entropy.

The basis of transhumanist "philosophy" relies on the hypothesis of an uninterrupted acceleration of technological means induced by the scientific progresses. This evolution supposes a genuine need for energy; this point is seldom evoked in the discussions of transhumanists but remains a fundamental requirement to support any important breakthrough, yet it might be easily observed and understood in modern time. Here, we are not dealing with philosophically-minded people but with trivial realities of the material world. Each time the new serendipity cumulated with the previous ones in order to accelerate and advance the "progress" of our evolution. This was not enough to drastically change the Human into the Transhuman, but they have assuredly contributed to bring us to where we are today.

Across the millennia, new kinds of Energy come into being, were classified, transformed and finally used. Every time these discoveries had a direct impact on our livelihood, our health, and our civilizations. From the beginning, Energy has behaved as a vector of the evolution of our civilizations.

In that respect it must be remembered that robots of any kind have voracious energy requirements, let alone the calorigenic data centers. Robots, wherever they are, fully depend on electricity as energy to function and operate, just as humans require food for survival, especially when deploying them for distant missions far away in space. Data centers are gobbling electricity especially quickly and the biggest ones need to be installed close to a large power station in order to limit transmission losses.

Energy is a prerequisite of our future whatever it may be. This ever-increasing thirst for energy is, by no means, a new phenomenon but persists throughout our civilization and has driven us to innovate and discover new sources of energy (and *vice versa*). This time we are desperately looking for a new kind of energy to sustain us in the 2.0 world. Augmented intelligence (artificial or not) is useless without the power to drive it.

Newton's laws of motion are the most fundamental in classical mechanics. Although these laws are accurate enough at very small scales and at low speeds, there are certain circumstances in which they are considered to be inappropriate in explaining the phenomena. Extremely high speeds (close to the speed of light) and very strong gravitational fields demand the use of special relativity and general relativity correspondingly. On the other hand, the laws of conservation of energy, momentum, and even angular momentum, are universal and can be applied to both light and matter, in classical and non-classical physics.

Sir Isaac Newton's first law: "Energy is not lost or destroyed: it is merely transferred from one party to the next." This means that you cannot make energy out of nothing. The second law refers to the state of energy and is reflected in a measurement of the degree of disorder (a measurement called entropy). In summary, when we use an energy source it is not destroyed but enters a more disordered state. This makes the energy less available to us, and converting the energy to power means some loss. As we have mentioned, the universe is winding down. The third law is that everything comes to a stop only when the temperature is at −273.15°C on the Celsius scale. This is called absolute zero and is where the entropy measurement is 0 (zero).

Together, these laws help lay the foundations of modern science. These laws are absolute physical laws — everything in the observable universe can be explained using them. Like time or gravity, nothing in the universe is exempt from these laws.

2. Empirical and scientific chemistry

Chemistry is the study of matter and the changes that matter undergoes. The properties of a substance may be quantitative (measured and expressed with a number) or qualitative (not involving numbers). A chemical property is a property that is determined only as the result of a chemical process. The original substance is converted to a different substance.

In chemistry, the empirical formula of a chemical compound is the simplest whole number ratio of atoms present in a compound. A simple example of this concept is that the empirical formula of sulfur monoxide, or SO, would simply be SO, as is the empirical formula of disulfur dioxide, S_2O_2. Thus, sulfur monoxide and disulfur dioxide, both compounds of sulfur and oxygen, have the same empirical formula. However, their molecular formulas, which express the number of atoms in each molecule of a chemical compound, are not the same.

The molecular formula, on the other hand, shows the number of each type of atom in a molecule. The structural formula shows the arrangement of the molecule. It is also possible for different types of compounds to have equal empirical formulas.

Traditionally, chemistry is classified into five main branches: Analytical chemistry involves the analysis of chemicals; Biochemistry uses chemistry to understand how biological systems work at a chemical level; Inorganic chemistry studies the chemical compounds in inorganic, or non-living things such as minerals and metals; Organic chemistry deals with chemical compounds that contain carbon, an element considered essential to life; Physical chemistry uses concepts from physics to understand how chemistry works.

Chemical energy is defined as energy released as a result of chemical reaction usually associated with generation of heat as a by-product, known as an exothermic reaction. The examples of stored chemical energy are biomass, batteries, natural gas, petroleum, and coal. For the most part, when chemical energy is released from a substance, it is transformed into an entirely new substance.

Chemical potential energy, such as the energy stored in fossil fuels, is the work of the Coulomb force during rearrangement of configurations of electrons and nuclei in atoms and molecules. Thermal energy usually has two components: the kinetic energy of random motions of particles and the potential energy of their configuration.

Another way to expand chemical energy is to put aside different chemical species endowed with different contact potentials including some kind of electrolyte. The invention of the modern battery is often attributed to Alessandro Volta (Figure 5.1). It actually began with a surprising accident involving the dissection of a frog. Volta hypothesized that the frog's leg impulses were actually caused by different metals soaked in a liquid. He repeated the experiment using cloth soaked in brine instead of a frog corpse, which resulted in a similar voltage. Volta published his findings in 1791 and later created the first battery, the voltaic pile, in 1800.

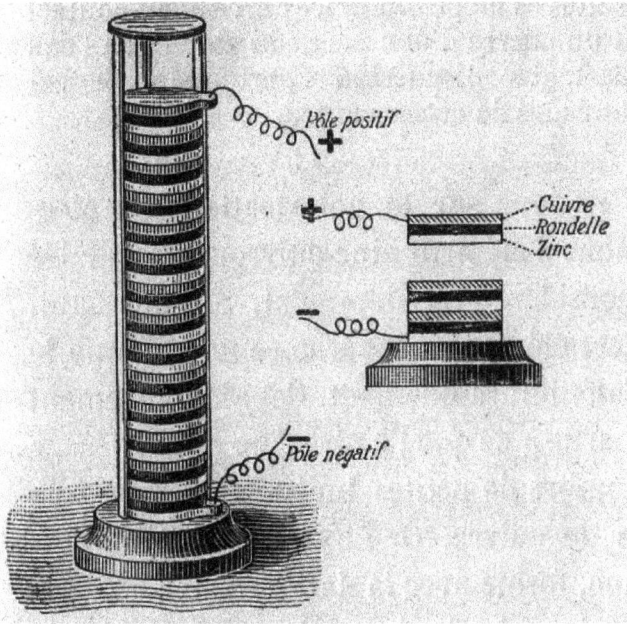

Figure 5.1.   A voltaic pile.

Now batteries have found their way into a wide spectrum of applications where electrical energy is needed to power appliances.

3. The Big Bang and the origin of things

One of the main questions we may ask is "where does everything come from?" The tentative answers differ widely depending on who was asking the question: from a physicist or religious person. In any case there is no definite answer to this question: the theories vary from purely imaginary constructs to arguments of physical research.

3.1. *Religion*

Religions are ideological structures elaborated by the human search for an explanation and a model of behavior to be followed by the faithful.

There have been, through millennia, a host of beliefs so popular and wide-spread yet afterwards became extinct; however some were more success-ful and are still widely practiced: Christianity, Judaism, or Islam, each with its own philosophy and requirements. Buddhism, a rather special case, is not founded on a notion of a Divine Being. Most of these religions assign a greater or lesser importance to the problem of the origins of the world.

They all imagine a superior authority (more or less) being an image of an old and venerable man (no other model was conceivable), provided with powers beyond imagining (and infinite energy). They call Him God, Jehovah or Allah and, of course, He is at the origin of everything. We have to go no further to find another explanation.

The same is true for the origin of life which obviously has to belong to this Entity. Here, every religion implicitly assumes an infinite power, that is to say a huge supply of energy, of whatever sort (as Jupiter, the Lord of Thunder (Figure 5.2)).

Figure 5.2.   Jupiter on Mount Olympus.

Some say it took six days to create the world and the seventh day was devoted to rest. Adam was created below an apple tree and Eve also to tempt him! Nonsense! Some believe the Big Bang implies a creator, while others argue that Big Bang cosmology makes the notion of a creator unnecessary.

3.2. *Science*

More rationality is required to get the beginning of an answer (as well as an answer for the beginning). Geologists together with astronomers and cosmologists tried to assemble observations and knowledge to hypothesize a theory supported by what they could observe, what they thought about the Big Bang: A theory of the instant zero of our world or "initial singularity". We shall try to get together some selected ideas taken in the literature, even if in a certain disorder.

Following this theory, on a certain day (though nobody can truly say when or where, because there was yet neither calendar nor map) a big explosion should have happened that came from nowhere, with an incredible quasi-infinite energy, throwing matter and radiation all over space (which is supposed to exist before?). From then, it became possible to make a coherent scenario. This is a quick and naïve view of the problem. Getting more closely is not obvious and requires a development.

As with any theory, a number of mysteries and problems arise as a result of the development of the Big Bang theory (Figure 5.3). Some of

Figure 5.3. The Big Bang created the Universe.

these mysteries and problems have been resolved while others are still outstanding. How the initial state of the universe originated is still an open question, but the Big Bang model does constrain some of its characteristics. While it is not known what could have preceded the hot, dense and compact state of the early universe, or how and why it originated, or even whether such questions are sensible, speculation abounds on the subject of "cosmogony". As such, physics may conclude that the notion of time did not exist before the Big Bang and since then it has been expanding and cooling down (which supposes that time has begun).

The earliest phases of the Big Bang are subject to much speculation, since astronomical data about them is not available. In the most common models the universe was filled homogeneously and isotropically with a very high energy density, extremely high temperature and pressure, and then expanded and cooled very quickly.

The period from 0 to 10^{-43} s into the expansion, the Planck epoch, was a phase in which the four fundamental forces — the electromagnetic force, the strong nuclear force, the weak nuclear force, and the gravitational force, were unified into a single phenomenon. At this stage, the characteristic scale length of the universe was the Planck length, 1.6×10^{-35} m, and consequently had a temperature of approximately 10^{32} K. Even the very concept of a particle breaks down in these conditions. A proper understanding of this period awaits the development of a theory of quantum gravity. The Planck epoch was succeeded by the grand unification epoch beginning at 10^{-43} s, where gravity separated from the other forces as the temperature of the universe fell.

After its initial expansion, an event that is by itself often called "the Big Bang", the universe cooled sufficiently to allow the formation of subatomic particles, and later atoms. Giant clouds of these primordial elements — mostly hydrogen, with some helium (and lithium) — later coalesced through gravity, forming early stars and galaxies, the descendants of which are visible today.

As the universe cooled, the remaining energy density of matter came to gravitationally dominate that of photon radiation. After about 379,000 years, the electrons and nuclei combined into atoms (mostly hydrogen), which were able to emit radiation. This remaining radiation, which continued through space largely unimpeded, is known as the cosmic microwave background (CMB) radiation.

The hot Big Bang predicted a uniform background radiation throughout the universe caused by the high temperatures and densities in the distant past. A wide range of empirical evidence strongly favors the Big Bang explanations, which is now essentially universally accepted.

The expansion of the Universe was inferred from early 20th century astronomical observations and is an essential ingredient of the Big Bang theory. Mathematically, general relativity describes space-time using the idea of the metric tensor, which determines the distances that separate nearby points. The points, which can be galaxies, stars, or other objects, are specified using a coordinate chart or "grid" that is laid down over all space-time. The cosmological principle implies that the metric tensor should be homogeneous and isotropic on large scales. In other words, the Big Bang is not an explosion *in space*, but rather an expansion *of space* (you will appreciate the difference!).

Measurements of the redshift–magnitude relation for type Ia supernovae indicate that the expansion of the universe has been accelerating since the universe was about half its present age. To explain this acceleration, general relativity requires that much of the energy in the universe consists of a component with large negative pressure, dubbed "dark energy".

Indirect evidence for dark matter comes from its gravitational influence on other matter, as no dark matter particles have been observed in laboratories. Many particle physics candidates for dark matter have been proposed, and several projects to detect them directly are underway.

One of the common misconceptions about the Big Bang model is that it fully explains the origin of the universe. However, the Big Bang model does not describe how energy, time and space were caused, but rather it describes the emergence of the present universe from an ultra-dense and high-temperature initial state. It is misleading to visualize the Big Bang by comparing its size to everyday objects. When the size of the universe at Big Bang is described, it refers to the size of the observable universe, and not the entire universe.

For several decades, the scientific community was divided between supporters of the Big Bang and the rival steady-state model which both offered explanations for the observed expansion, but the steady-state model stipulated an eternal universe in contrast to the Big Bang's finite age; in 1964, CMB radiation was discovered, which convinced many cosmologists that the steady-state theory was disproved, since, unlike the steady-state theory.

After its initial expansion, an event that is by itself often called "the Big Bang", the universe cooled sufficiently to allow the formation of subatomic particles, and later atoms. Giant clouds of these primordial elements (mostly hydrogen, with some helium and lithium) later coalesced through gravity, forming early stars and galaxies, the descendants of which are visible today. This is a self-transforming energy process.

"The Big Bang" as an event is also colloquially referred to as the "birth" of our universe since it represents the point in history where the universe can be verified to have entered into a regime where the laws of physics as we understand them (specifically general relativity and the Standard Model of particle physics) work. Based on measurements of the expansion using Type Ia supernovae and measurements of temperature fluctuations in the cosmic microwave background, the time that has passed since that event — known as the "age of the universe" — is 13.8 billion years.

Besides these primordial building materials, astronomers observe the gravitational effects of an unknown dark matter surrounding galaxies. Most of the gravitational potential in the universe seem to be in this form, while the Big Bang theory and various observations indicate that this excess gravitational potential is not created by baryonic matter, otherwise known as the atoms encountered in normal research. Measurements of the redshifts of supernovae indicate that the expansion of the universe is accelerating, an observation attributed to the existence of dark energy.[1]

If you peer out into the depths of space — at the vast expanse of stars, galaxies, and even the leftover glow from the Big Bang itself — you might think that if humanity could understand the laws of nature and create a good enough technology, there are no limits to what we can explore. Nothing we can learn from simulations on the ground is as good as analyzing the observatory when it is up and running.

Our understanding of the universe back to very early times suggests that there is a past horizon, though in practice our view is also limited by the opacity of the universe at early times. So, our view cannot extend further backward in time, though the horizon recedes in space. If the expansion of the universe continues to accelerate, there is a future horizon as well.

Whether by chasing optical and ultraviolet light like Hubble or infrared light like Webb, telescopes can see further and more clearly when operating above Earth's distorting atmosphere. The telescope enables astronomers to peer back further in time than ever before, all the way

---

[1] Peebles, P. J. E. & Ratra, B. (2003). The cosmological constant and dark energy. *Reviews of Modern Physics*, *75*(2), 559.

back to when the first stars and galaxies were forming 13.7 billion years ago. That's a mere 100 million years from the Big Bang, when the universe was created. NASA's James Webb Space Telescope, the agency's successor to the famous Hubble telescope, (which orbits 330 miles, i.e., 530 km up), launched on December 25, 2021 on a mission to study the earliest stars and peer back farther than ever before into the universe's past.

The James Webb Space Telescope has finally arrived at its new home. The new space observatory reached its final destination, a spot known as L2. Technically known as the second Earth–Sun Lagrange point, it is a spot about 1.5 million kilometers from Earth in the direction of Mars, where the Sun and Earth's gravity balance out the inward-pulling centripetal force that keeps a smaller object on a curved path. That lets objects at Lagrange points (Figure 5.4) stay put without much effort. Pairs of massive objects in space have five such Lagrange points.

Lagrange points are positions in space where objects sent there tend to stay put. At Lagrange points, the gravitational pull of two large masses precisely equals the centripetal force required for a small object to move with them. These points in space can be used by spacecraft to reduce fuel consumption needed to remain in position.

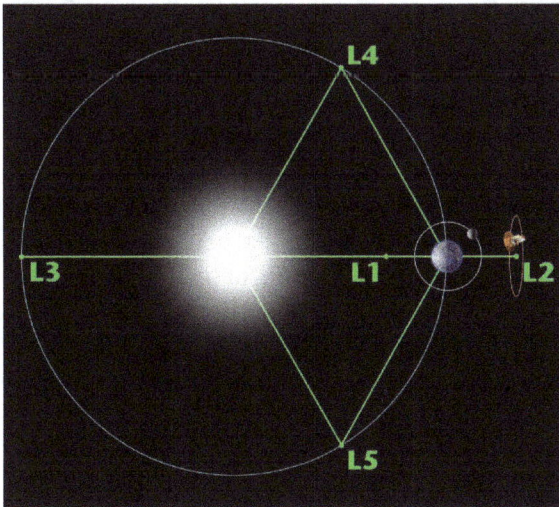

Figure 5.4. Organization of Lagrange points.

The amount of fuel needed to maintain Webb's home in space will set the lifetime of the mission. Once the telescope runs out of fuel, the mission is over. Luckily, the spacecraft had a near-perfect launch and didn't use much fuel in transit to L2. As a result, it might be able to last more than 10 years, team members say, longer than the original 5- to 10-year estimate.

Webb sees in infrared light, wavelengths longer than what the human eye can see. But humans do experience infrared radiation as heat. That means that the parts of the telescope that observe the sky have to be at about 40 K (−233°C), which nearly matches the cold of space. That way, Webb avoids emitting more heat than the distant sources in the universe that the telescope will be observing, preventing it from obscuring them from view.

3.3. *The origin of life*

The creation of our world is the first mystery of the story; the second is the following emergence of life leading to a thinking and feeling being. Where might that have come from? Here too, religions offer an explanation (God!) but without any scientific basis.

Although scientists cannot directly address how life begin on Earth, they can formulate and test hypotheses about natural processes that could account for various aspects of life given the geological evidence. In the 1920s, Alexander Oparin and J. B. S. Haldane independently proposed nearly identical hypotheses for how life originated on Earth. Their hypothesis is now called the Oparin-Haldane hypothesis.[2]

They reasoned that atmospheric oxygen sustained the synthesis of organic molecules. Organic molecules are the necessary building blocks for the evolution of life. In his *The Origin of Life*, Oparin[3] argued that a "primordial soup" of organic molecules could be created in an oxygen-less atmosphere through the action of sunlight. These would combine in ever-more complex fashions until they formed droplets. These droplets would "grow" by fusion with other droplets, and "reproduce" through fission into daughter droplets, and so have a primitive metabolism in which

---

[2] Oparin A. I. (1924). *Proiskhozhozhdenie zhizny*. Moscow (translated by Ann Synge, in Bernal 1967. *The Origin of Life*. Weidenfeld and Nicolson, London.
[3] Oparin A. I. (1952). *The Origin of Life*. Dover, New York.

those factors which promote "cell integrity" to survive for those that do not become extinct. Many modern theories of the origin of life still accept Oparin's ideas as a starting point.

The environment that existed in the Hadean Eon was hostile to life, but how much so is not known. There was a time, between 3.8 and 4.1 billion years ago, which is known as the Late Heavy Bombardment. It is so named because many lunar craters are thought to have formed then. The situation on other planets, such as Earth, Venus, Mercury and Mars must have been similar. These impacts would likely sterilize the Earth (kill all life), if it existed at that time.

But it was not until 1953 that these ideas received experimental support. Stanley Miller, then a graduate student working in the laboratory of Harold Urey, set up his famous experiment in which electrical discharges were passed through a mixture of gases (methane, ammonia, hydrogen, and water vapor) simulating a thunderstorm on the primitive Earth. The experiment produced a mixture of several amino acids, the building blocks of proteins. Miller speculated that this was how organic compounds had been made on the early Earth. In the same year, Crick and Watson published their structure for DNA, the first step in elucidating the fundamental molecular basis of life. These two discoveries meant that we had both a plausible way of generating simple organic building blocks and an understanding of the macromolecules on which life depends. A detailed experimental and theoretical study of the origin of life was now possible.

If life evolved in the deep ocean, near a hydrothermal vent, it could have originated as early as 4 to 4.2 billion years ago. If, on the other hand, life originated at the surface of the planet, a common consensus is it could only have done so between 3.5 and 4 billion years ago.[4]

Originally, the atmosphere of Earth had almost no free oxygen. It gradually changed to what it is today, over a very long time (see Great Oxygenation Event). The process began with cyanobacteria. They were the first organisms to make free oxygen by photosynthesis. Most organisms today need oxygen for their metabolism; only a few can use other sources for respiration. so it is believed that the first proto-organisms were chemo-autotrophs, and did not use aerobic respiration. They were anaerobic.

---

[4]Maher, K. A. & Stevenson, D. J. (1988). Impact frustration of the origin of life. *Nature*, *331*(6157), 612–614. https://doi.org/10.1038/331612a0.

Pre-cellular systems called liposomes have been identified in which the encapsulation of DNA is achieved using dehydration-hydration cycles similar to those that may have occurred in an intertidal setting on the early Earth.[5] Once self-replication of life was established, then evolution to more complicated forms of life could have taken place by Darwinian selection.[6] The ocean could have played further roles in human development, for example via the aquatic ape hypothesis, although that particular theory is not widely accepted.

About three billion years ago, something similar to modern photosynthesis developed and oxygen was produced. Over time, it transformed Earth's atmosphere to its current state. Some of the oxygen reacted to form ozone, which collected in a layer near the upper part of the atmosphere. By blocking the ultraviolet radiation, it allowed cells to colonize the surface of the ocean and ultimately the land. Fish, the earliest vertebrates, evolved in the oceans around 530 million years ago.

In the 1960s, the hydroxyl radical (OH) was identified by astronomers in these interstellar clouds, as was ammonia (NH_3). Later, water (H_2O) was identified soon to be followed by the important organic molecule formaldehyde (H_2CO). At the beginning of this century there have been over 100 organic molecules identified in space. These organic molecules were not only identified within our own Milky Way Galaxy, but also in other galaxies. So, it looks like the existence of organic molecules is a universal phenomenon, and is driven by the energy from stars and follows the laws of chemistry.

The first step in life was to build complex biomolecules and this is accomplished by extracting energy from the surrounding environment. But now we have much evidence from spectroscopy indicating that these complex biomolecules were formed in interstellar clouds of dust and gas. The same dust and gas clouds, light years across, where stars and planets form.

---

[5] Oró, J., Miller, S. L., & Lazcano, A. (1990). The origin and early evolution of life on Earth. *Annual Review of Earth and Planetary Sciences, 18*, 317–356. https://doi.org/10.1146/annurev.ea.18.050190.001533.

[6] Benn, C. (2001). The Moon and the origin of life. *Earth, Moon and Planets, 85/86* (61–66). https://doi.org/10.48550/arXiv.astro-ph/0112399.

4. Science and longevity

Knowledge is the key element that differentiates Men from animals. In the past, knowledge was the preserve of a few individuals driven by the need to understand the surroundings, the weather, and everything related to life and survival. This knowledge was generalized into concepts, developed into ideas, classified into ideas which were kept related through well-founded logical linkages. Currently, that thirst for knowing has spread widely, combined with the idea that it is a source of wellbeing and daily comfort. That is the only remaining trace, the building block, even if small. The proportion of people involved increasingly evolves, but a long way still remains in front of us. One can dream of a future humanity fully devoted to research enriching our knowledge, with help from intelligent technology!

All along Darwinian human evolution, Energy has followed, preceded, motivated, even initiated and accelerated the successive steps of civilizations. The Bronze Age, Iron Age, Steam Machine Age, Electricity Age, Nuclear Age and now Digital Age have all been powered by increasing available energy, with the longevity of men extended correspondingly. Now the 2.0 world has to face some new issues related to the expanding and inevitable vital need of energy. There are now the key challenges for humanity to survive.

The accumulation of our collective knowledge we call science, has come into being as we know today, during this new 2.0 century, a level so exceptional that this plays out on our intimate achievement: our statistical longevity is increasing at an impressive rate (Figure 5.5).

This is not without producing deep consequences in the way our societies evolve. Various factors contribute to an individual's longevity. Significant factors in life expectancy include gender, genetics, access to health care, hygiene, diet and nutrition, exercise, and lifestyle. In pre-industrial times, deaths at young and middle age were more common than they are today. This was not due to genetics, but because of environmental factors such as disease, accidents, and malnutrition, especially since the former were not generally treatable with premodern medicine. Deaths from childbirth were common for women, and many children did not live past infancy. In addition, most people who did attain old age were likely to die quickly from the above-mentioned untreatable health problems.

There is debate as to whether the pursuit of longevity is a worthwhile health care goal. Bioethicist Ezekiel Emanuel, one of the architects of

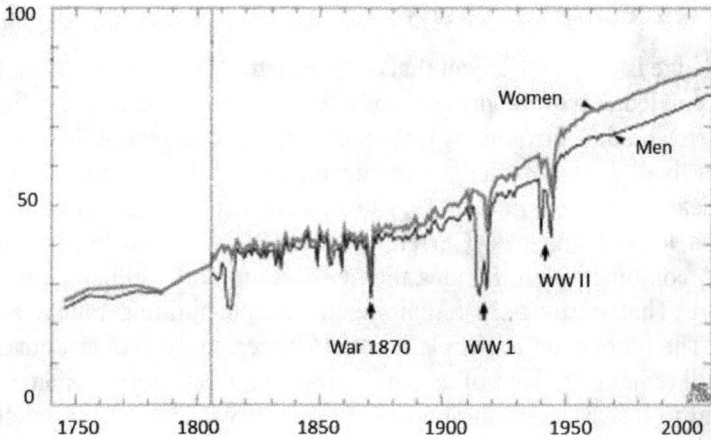

Figure 5.5.   Mean life expectancy in Europe in recent times.

Obama-Care, has argued that the pursuit of longevity via the compression of morbidity explanation is a "fantasy" and that longevity past age 75 should not be considered an end in itself. The United Nations has also made projections far out into the future, up to 2300, at which point it projects that life expectancies in most developed countries will be between 100 and 106 years and still rising, though more and more slowly than before.

However recent increases in the rates of lifestyle diseases, such as obesity, diabetes, hypertension, and heart disease, may eventually slow or reverse this trend toward increasing life expectancy in the developed world, but have not yet done so. All of that implies a large increase of the energy required to sustain these new populations (food, medical cares, housing, etc.).

5. Extrapolations

Among the various sources of energy available on Earth, some are easily accessible such as conventional fossil fuels or coals, and others come from natural phenomena that are directly accessible (waterfalls, dams on a river, geothermal, etc.). Of course, we must also mention the solar radiation which is present in almost every aspect of our life process. However, this energy requires some effort to be converted into something useful for us. Solar cells are the most prevalent converters to provide electricity. Nevertheless, the sun's energy is not permanent, as its supply alternates between night and day, thus requiring a buffer system such as batteries, which again lowers the final yield. An elegant solution would be to place solar collectors in orbit so that they receive the solar light all the time; but the trick is to bring this energy down to earth!

The wind is also considered as a source of renewable energy generated by adapted gigantic (and costly) wind turbines, and again, some form of storage facility is also required. Here too, the final yield keeps being severely impacted.

These intermittent (Figure 5.6) renewable solutions merely compensate for the occasional shortfalls from the traditional sources, and they cannot suffice to be considered as large enough to be fully competitive.

However, the continuous rise in energy needs of our growing modern societies 2.0, leads to an active research of other solutions for a large-scale production, namely of electricity which remains the main vector to satisfy the fixed requirements.

Figure 5.6.   Combined solar cell and wind turbine plants.

The known best solution, rain or shine, to provide KWh in abundance remains ineluctably the atom nucleus. Fission reactors plants abound all over the world. At present, the vast majority of electricity from nuclear power is produced by nuclear fission of uranium and plutonium in nuclear power plants. There are 444 civilian fission reactors in the world, with a combined electrical capacity of 396 Gigawatts (GW). There are also 53 nuclear power reactors under construction and 98 reactors planned, with a combined capacity of 60 GW and 103 GW, respectively. This is a lot. They are the result of a long scientific history that will be pursued by the knowledge of the fusion phenomenon not yet fully mastered with the many experimental "Tokamak" settled in Europe, the US, Russia, Korea, or China.

Research into using fusion for the production of electricity has been pursued for over 60 years. Although controlled fusion is generally manageable with current technology (e.g., fusors), successful accomplishment of economic fusion has been stymied by scientific and technological difficulties, nonetheless, important progresses have been made. At present, controlled fusion reactions have been unable to break-even (self-sustaining). The two most advanced approaches are magnetic confinement (toroid designs) and inertial confinement (laser designs).

The release of energy with the fusion of light elements is due to the interplay of two opposing forces: the nuclear force, which combines together protons and neutrons, and the Coulomb force, which causes protons to repel each other. Protons are positively charged and repel each other by the Coulomb force, but they can nonetheless stick together, demonstrating the existence of another, short-range, force referred to as nuclear attraction.

The Sun is a main-sequence star, and, as such, generates its energy by nuclear fusion of hydrogen nuclei into helium. At its core, the Sun fuses 620 million metric tons of hydrogen and makes 616 million metric tons of helium each second.

Energy released in most nuclear reactions is much larger than in chemical reactions, because the binding energy that holds a nucleus together is greater than the energy that holds electrons to a nucleus. For example, the ionization energy gained by adding an electron to a hydrogen nucleus is 13.6 eV — less than one-millionth of the 17.6 MeV released in the deuterium–tritium (D–T) reaction. Fusion reactions have an energy density many times greater than nuclear fission.

Figure 5.7. Nuclear tokamak.

Confinement refers to all the conditions necessary to keep a plasma dense and hold the high temperature long enough to undergo fusion. However, the necessary combination of temperature, pressure, and duration has proven to be difficult to break-even practically and economically. Research into fusion reactors began in the 1940s, but to date, no design has yet produced more fusion power output than the electrical power input. A second issue that affects common reactions is managing neutrons that are released during the reaction, which over time degrade many common materials used within the reaction chamber.

Fusion researchers have investigated various confinement concepts. The early emphasis was on three main systems: z-pinch, stellarator, and magnetic mirror. The current leading designs are the tokamak (Figure 5.7) and inertial confinement fusion (ICF) by laser. Both designs are under research on very large scales, most notably the ITER tokamak in France, and the National Ignition Facility (NIF) laser in the United States. Researchers are also studying other designs that may offer cheaper solutions. Among these alternatives, there is an increasing interest in magnetized target fusion and inertial electrostatic confinement, and new variations of the stellarator.

Beside the efforts to implement new sources of clean energy, we are also are making huge effort to save energy when operating industrial machines as well as a variety of household appliances such as washing machines, television (TV) sets, ovens and so on. No need to mention the care taken with private cars to make them more gas-efficient and less polluting. This has led to the development of electrically powered vehicles, which are both clean and economic to use, but with the drawback of the need to carry heavy and expensive batteries.

A balance has to be struck between the limited sources of energy and our increasing need to sustain the requirements of the modern world.

In the earliest times, Men were at the at the mercy of the power of Nature. Now we have developed and built a huge technological world, but the following and currently looming need is to gather enough energy to operate the machines supporting our civilizations.

Chapter 6

Civilizations and Induced Issues

Civilizations result from the pooling of technological knowledge as well as culture and language in a new society. This change originates in an impulse arising from a new set of opportunities (trade, resources, religion, technological means or else). However, to be triggered, they need a significant amount of input, that is to say, Energy.

As a matter of fact, war is often the source that triggers the sequence of events to disturb the stability of the equilibrium, but at the same time, paving the way for the arrival of the new order just like a phoenix rising from the fire. Athens as well as Rome has suffered the impact of invaders and reciprocally, their lifestyle has spread outside. All this occurs with significant energy expenditure following a random process. Things do not happen on their own! It is often required to destroy the existing old world order before building a new one.

1. The evolution

Every evolution in our societies relies on a peculiar way to take advantage of an energy source. This is the driving force behind our civilization. This is especially true during the time of industrial civilizations which are energy-hungry by nature. In the old days, the energy was basically muscle based, provided by slaves, and animals, such as horses, camels, even elephants. So, new structures such as the Egyptian pyramids were built from the ground up as impressive achievements; we could never do that again, but the true global change began with the machines that opened up a new dimension driven by new forms of energy.

Machines driven by all kinds of energy made it possible to tackle new issues, solve or improve the functionality of some forms of machine which powered human civilizations from stage to stage. We call this evolution "progress" during which new tools produced new behaviors, consumptions, habits, and forms of communication which led us to the 2.0 World, marked by the arrival of Artificial Intelligence (AI).

This AI progresses at an incredible rate to pilot cleverly any operation, using logical deductions (even play Go). However, for time being (hopefully), the intelligent human brain, (yes, there are some!) remains in control of the game. Looking at the game of war led by one man playing out in Eastern Europe, this assuredly might be well more dangerous than an AI machine. We cannot be certain that someday a crazy dictator or president, might not press a red button and bring the apocalypse on us. The energy implied should be enormous, at a global scale, including the replies.

Apart from these sad perspectives of AI evolution into every aspect of our life, from biology, medicine, to computer sciences, there has been more exciting progress in human longevity (some dreamers are even claiming immortality). We can all clearly see this in the obituary columns of the daily newspapers showing an increasing number of centenarians. Natural death no longer exists, one dies always of a determining cause, even if some scientists emphasize to "reversing death" in all seriousness!

"Transhumanism (Figure 6.1) is an international philosophical movement that advocates for the transformation of the human condition by developing and making widely available sophisticated technologies to greatly enhance human intellect and physiology".[1]

From Gilgamesh to the *Übermensch* of Friedrich Nietzsche there has been a recurrent dream of what could we could do to gain access to a different (or better) being. Transhumanism[2] can be considered as the culmination of a continuous evolution from the Big Bang. Darwinian theory says that genetic evolution occurs with the millenaries, which serves to make the species more closely adapted to the requirements of the environment. All of these follows a continuous use of energy driven by the arrival

---

[1]Bostrom, N. (2005). A history of transhumanist thought. *Journal of Evolution and Technology*, *14*(1), 1–25.
[2]Fillard, J. P. (2020). *Transhumanism: A Realistic Future?*. World Scientific Publishing Company.

Scientists Use Stem Cells to Reverse Death
Could we actually be one day closer to scientifically reversing death? We most definitely could be with study expected to be launched later this year

Figure 6.1. The transhumanist hope for immortality.

of new nanotechnology, biotechnology, information technology and cognitive technology (NBIC technologies).[3]

Human longevity has already grown to the point that some believe it could be possible to reverse the process of aging and bring up the possibility of a rejuvenescence that could benefit transhumans (or perhaps, posthumans). We also imagine soon being able to create life from scratch, although this looks to be beyond our current capabilities.

The difficulty shall be solved for a peaceful purpose: energy. From ancient times until now, a new kind of energy has been the trigger for new civilizations: now we have entered into Civilization 2.0, which could announce the future transhumanism, but not a new source of energy.

2. Old civilizations

The evolution of different world civilizations followed different paths; some were fast-tracked, while others progressed organically. The Aboriginal peoples in the awesome expanse of Australia, have remained

---

[3]Nanotechnologies, Biotechnologies, Information technologies, Cognitive sciences.

Figure 6.2.   This ziggurat was built without any wheelbarrows.

outside of the main intermingling of populations. They knew how to make fire and invented the boomerang but did not build any lasting edifices.

The Inca and Maya (Figure 6.2) did not invent the wheel but that did not prevent them building stone edifices and pyramids still standing today. Egyptians too have built impressive pyramids with the hands of slaves. They all belong to structured civilizations and cultures.

The Chinese were by far the more creative minds, with scripture, bit and bridle, black powder, the rudder and so on. Yet they did not exploit these inventions as other people later would do. Each civilization has its own way of doing things.

The Mesopotamian civilizations benefited from trade exchanges by caravans between East and West (Silk Road) and initiated intellectual progresses, that led to the European Middle Ages, a period when fortresses and castles sprang up everywhere. In every place, each problem was solved with a particular solution and an adapted use of the energy found in a diversity of sources (animal energy, vapor machines, fuel, electricity and so on).

Figure 6.3.   Hiroshima, August 6, 1945.

2.1. *War*

War also was part of "progress" and largely threatens us with very high energy weapons. Hopefully they are so frightening that nobody dares use them except as a threat which is counterbalanced by a possible reply at the same level, which is crazy. This was named the "balance of terror".

Fortunately, such a balance has worked for now. The bombs of Hiroshima and Nagasaki (Figure 6.3) are still present in our collective memories. That did not prevent the proliferation of many powerful weapons. These first atomic bombs (A-bombs) were driven by uranium (Little Boy) or plutonium (Fat Man) fission kernels and developed some 23 kilotons (TNT) of energy. The fire bowl was 180 m in diameter.

The most powerful hydrogen bomb (H-bomb) (Tsar Bomba) today is wildly disproportionate in comparison with prior ones. The fire bowl would be 12 km in diameter, which means that a city like Paris would be completely vitrified in a single shot and the total destruction zone would extend a diameter of 150 km. The H-bomb uses a first stage explosion of a A-bomb as a starter to light up a fusion reaction. That gives a terrifying mushroom!

3. Present civilization 2.0

Now as we enter into the 21st century, we have inherited a huge diversity of energy sources, from nuclear power to the battery. Our civilization has become a worldwide enterprise. And, once again, the transistor has given us increasingly sophisticated means to power almost everything in life or work to the extent that it has transformed our way of life, and to put decision making and control into many applications we use today. These transistors must be powered by electricity generated by power stations or batteries.

There is a wide variety of rechargeable or non-rechargeable (alkaline) batteries mostly using lithium ions as an electrolyte between an anode (graphite) and a cathode (metal oxide). The size (and corresponding) storage capacity of these devices vary from little buttons for watches or other mobile devices to much larger (and more expensive) ones, for instance the batteries to power a car or even a submarine.

Batteries have become (Figure 6.4) more and more part of our lives, that is to say, the market is burgeoning and the key basic ingredients ever more difficult to find. The conclusion remains that they will become increasingly expensive; keeping in mind that they have a limited storage capacity and life.

Figure 6.4.   Batteries in electric cars are heavy and bulky (and expensive).

3.1. *Computers*

Concerning computers, the size and energy consumption vary to a very large extent; from the isolated chips to the big data center, there is a huge diversity. Current robots all encounter a tremendous limitation on their autonomous life determined by the source of energy. Those which imitate humans or animals in the execution of a task, such as BigDog, must spend a great deal of energy to merrily carry a heavy battery with them. The harder they work, the less freely they can operate, and when the battery is empty the robot becomes a simple material object without a brain and without "life." However, unlike their human equivalent (namely concerning the brain), the robot is not dead — it just needs to be recharged to bring back their strength and intelligence; this is not the case for a human, for if he stays too long without food or air, he dies!

The Internet (Figure 6.5) is a particularly huge consumer of energy, and this consumption is growing steadily. The work capacity of these

Figure 6.5. Microsoft data center.

centers is directly related to the many hard disks involved, and obviously their number increases constantly. The computing power of these centers will ineluctably be limited by the electrical power required and the corresponding costs, even if the specifications of the memories are constantly improving.

In the US, big data centers are mostly installed close to an electrical plant to reduce transmission losses; in Virginia, abandoned coal mines have been reopened, thus reviving a lost business but also reviving pollution. How long will we be able to tap on the natural resources with an ever-rising world population?

In France, where free space is more limited, data centers are often located at old disused industrial sites, close to downstream users — that is to say, inside the big towns, which produces complex connection problems with the grid so as to ensure sufficient power. Data centers are made of a huge collection of hundreds or thousands of hard disks (now solid-state drives (SSDs)) to keep pace with demand.

Between the computer and the brain, the availability of energy sources will certainly become a crucial technological issue. The Internet cannot grow indefinitely; all our energy resources cannot be only dedicated to feeding its hard disks, whatever the technological advances that are to occur. The only solution in the future (if any) could rely on the biological computer combining the sobriety of the biological cell and the efficiency of the transistor with the aim of obtaining better compliance with the brain. But a biological Internet is certainly not arriving tomorrow, and even a way to "feed" it is yet to be found.

Renewable energies (wind turbines as well as solar cells) are intermittent, unpredictable, and often unavailable (Figure 6.6) precisely when they are needed; in any case, they all require a reliable complementary source to back them up at any moment.

The intelligent computer, of whatever kind (implant, android, cyborg, or connected) will be facing not an individual human but assembled societies of which it is the product and which feed it energy. It will depend on them no matter what it can do, so long as it is unable to fully support itself. Apart from its possible high intelligence, it will have to really become as efficient and versatile as humans are. Whatever their performance, computers, as of now, will not be competitors with human brains in the domain of intuition, imagination or creativity; they are restricted to logical operations.

Figure 6.6. Even wind turbine blades can freeze.

4. Space colonization

The increasing development of space technology, knowledge and facilities has led to a substantial occupation of the space around the Earth with satellites that provide us with many facilities to keep up our lifestyle, such as Global Positioning System (GPS) to find our way around. Even permanent living in space stations has become a common topic today. This led us to the idea of going further: why not bluntly emphasize changing the boat and migrate to another planet which could be hospitable for human life? If such a planet does exist, would it be ready to accommodate us?

Many arguments have been made for and against space colonization. The two most common in favor of colonization are survival of human civilization and the biosphere in the event of a planetary-scale disaster (natural or man-made), and the availability of additional resources in space that could enable expansion of human society. The most common objections for colonization include concerns on the common modification of the cosmos may be likely to enhance the interests of the already powerful (including major economic and military) institutions; that could produce an enormous opportunity cost as compared to expending the same resources here on Earth, as well as an exacerbation of the pre-existing detrimental processes such as wars, economic inequality, and environmental degradation.

To date, no space colony has been established. A space colony would set a precedent that would raise numerous socio-political questions. The mere construction of the needed infrastructure presents a daunting set of technological and economic challenges. Space colonies are generally conceived as organizational and material structures that have to provide for nearly all (or all) the needs of larger numbers of humans, in an environment out in space that is very hostile to human life and quite inaccessible to maintenance and supply from Earth.

Many proposals, speculations, and designs for space settlements have been put forwarded, and a considerable number of space colonization advocates and groups are actively advocating their Utopias. Several famous scientists, such as Freeman Dyson, have come out in favor of space settlement. The first idea is colonization of our nearest neighbor: the Moon. But what is there to do on the Moon's surface apart from revolving around the Earth? There is no usable atmosphere, no radiation, no water, no precious minerals, but plenty of solar energy!

Then, other more remote exo-planets in our galaxy (some 3,000 following Kepler's mission analysis) can be emphasized but the nearest one is located some 12 light-years from us! Apart from fully livable planets, some "moons" can be found, around Jupiter for example (Figure 6.7).

In any case, such a transfer of inhabitants would require special rockets possessing substantial resources of energy in various forms: oxygen,

Figure 6.7.   10 "moons" orbiting Jupiter!

food, radiation protection, equipment, batteries and solar converters in addition to rocket fuel.

5. Aliens and so on

Another way to approach the problem is to ask ourselves a question: are we alone in the universe? What kind of reply would you expect? Are there similar life forms somewhere in the universe? Would they be more advanced than us or would they be far behind us?

Many attempts have been made (in vain) to meet with alien intelligence: radio messages, far-off satellites, telescopes, etc. no real reply to date! If they exist somewhere, would they be able to understand the message? What kind of reply should we expect? Would they be kind enough to give a reply? What would be their level of development, would they be ahead of us or behind? It took us thousands of millennia to get to our current stage.

However, there are reported sightings of unidentified flying objects (UFOs) clouded in mystery mostly flying at night. These objects display strange behaviors in open defiance of gravity and known laws of physics (not to mention the concept of energy). Obviously, a clear distinction with fake views is not so easy to reach! Could these objects (Figure 6.8) belong to another planet and another lifeform that under conditions different from ours?

Figure 6.8. Supposed alien and UFO!

From an "alien autopsy" video dating back to 1995 to a video of fighter jets apparently chasing a UFO, there are hundreds of videos that people have claimed to be proof of real alien life. Hence, the question arises: is this real or fake?

The Roswell UFO incident is a world-renowned conspiracy theory about the existence of aliens. In the summer of 1947, a farmer discovered unidentifiable debris in his sheep fields just outside Roswell, New Mexico. A local Air Force base claimed the debris came from a crashed hot air balloon, but many people reckon they are really the remains of a flying saucer!

In April 2018, it was reported that Buzz Aldrin, along with three other astronauts, had passed lie detector tests over claims they had experienced alien encounters. Aldrin, now 88, claimed he saw a spaceship on his way to the moon, adding: "There was something out there that was close enough to be observed... sort of L-shaped." The tests, reportedly more reliable than standard lie detector tests, showed he was telling the truth.

So, could there be another way to look at energy in other civilizations using a different chemistry? Could silicon be a credible alternative to carbon? One thing is for certain: the chemical atoms and species and the local conditions must be the same whatever the place where they are located. Would it be possible to live in an environment without gravity and inertia? What kind of new force field would be involved? What could be the implication of the time parameter in the explanation of these phenomena of quick movements and vanishing.

If such alien and highly developed civilizations do exist, why are they keeping such a low profile until now; do they fear us as we fear wild beasts? Would they be sensitive to viruses as it was suggested by George Orwell?[4] Would they consider us stupid animals? Are they afraid of us because of our tendency to wage war? A potential flaw is assuming an extraterrestrial civilization would be only a few hundred years ahead of us in technology. How about civilizations that may be a million years ahead of ours? One thing is for sure: they would need energy of some form to survive.

In 1964, Russian astrophysicist Nikolai Kardashev categorized civilizations into three types based on their energy output.[5] Now, we have

---

[4]Orwell, G. (1949). *1984*. Alma Classic Evergreens 2021.

[5]Kardashev, N.S. (1964). Transmission of information by extraterrestrial civilizations. *Soviet Astronomy*, **8**, 217–221.

added three more levels based on mathematics and theoretical physics. As the levels increase, the abilities of the civilizations sound like science fiction.[6]

Energy is the only universal currency; it is necessary for getting anything done. The conversion of energy on Earth ranges from the terraforming forces of plate tectonics to the cumulative erosive effects of raindrops. Life on Earth depends on the photosynthetic conversion of solar energy into plant biomass. Humans have come to rely on many more energy flows — ranging from fossil fuels to photovoltaic to generate electricity — for their civilized existence.[7] Humans (in our current opinion) are the only species that can systematically harness energies outside their bodies, using the power of their intellect and an enormous variety of artifacts.

6. Civilizations, progress and wars

It remains not to be forgotten that war has been, for centuries, a powerful stimulus for technical innovation and progress driven by energy. As stated by Bill Gates himself: "The past 300 years have seen the most miraculous advances in the human condition—and just about all of those advances can be traced directly to the exploitation of new forms of energy". It is fascinating to reflect on how much energy innovation occurred during the course of a single century in spite of (or because of) two bloody world wars.

With the arrival of the top technological Civilization 2.0, one could have thought that war definitively belongs to the past and that rationality has definitely prevailed. Nuclear weapons exist for calming things down. We were completely wrong; war has come again with its huge energy waste, and, to date, it did not bring any new technological bliss!

One of the things that we have increasingly been confronted with and have fought to both survive and eradicate in centuries past is the scourge of war between communities, including civilized communities. In some ways this might seem a bit at odds with the ideas of civilization, progress, and human perfectibility, but just as there is a close relationship between civilization and progress, so too there is a close relationship

---

[6]Asimov, I. (1979). *Extraterrestrial Civilizations*. Crown.
[7]Smil, V. (2017). *Energy and Civilization: A History*. The MIT Press.

between civilization and war, and between war and progress. Is war an essential part of the mechanism of evolution?

Clearly the twin ideals of civilization and progress are two important driving forces in our attempts to make sense of life through the articulation of an all-encompassing, or at least wide-reaching, philosophy of history. Indeed, in recent centuries it has proved irresistible to a diverse range of thinkers from across the political spectrum.

We generally say that today's human civilization has reached a very good stage in technology. But to classify these civilizations Nikolai Kardashev (a Russian astrophysicist) proposed a scale of three levels based on their energy harvesting and consuming capabilities:

- Type 1: refers to a civilization that could harness all kinds of energy on a planet with 100% efficiency. That means some 10^{16}–10^{17} watts.
- Type 2: refers to a civilization that would be able to harness the complete energy of their solar system (hypothetical structures like the "Dyson sphere"): collecting black holes to reach some 10^{26} watts.
- Type 3: refers to a civilization that would be able to harness the energy of their whole galaxy, that is to say, some 10^{37} watts.

But a big question remains that if such civilizations really exist why did they not meet us or invade us up until now?

In recent years, the theory of "Dialogue Among Civilizations", a response to Huntington's "Clash of Civilizations",[8] has become the center of some international attention. Huntington argues that the trends of global conflict after the end of the Cold War are increasingly appearing at these civilizational divisions. Huntington believed that while the age of ideology had ended, the world had only reverted to a normal state of affairs characterized by cultural conflict. He writes: "This is not to advocate the desirability of conflicts between civilizations. It is to set forth descriptive hypothesis as to what the future may be like".

Why do people go to war? Is it rooted in human nature or is it a late cultural invention? How does war relate to the other fundamental developments in the history of mankind? With the advent of the industrial revolution and subsequent social change, the dynamic, the politics, and the economics of war changed: wealth and resources increased, individuals

---

[8] Huntington, S. (1993). The clash of civilizations?. *Foreign Affairs*, *72*(3), 22–49. https://doi.org/10.2307%2F20045621.

became more invested in home and business. Among those societies that have not benefited by social and economic change, the drivers remain and change to the status quo is often actively sought. Therein lies the risk of current and future conflict, which would be all the more dangerous because of our modern capabilities. Indeed, as he implies in his last sentence, the final history of warfare has yet to be written.

Our Earth is 4.5 billion years old, but humans evolved only around 200,000 years ago. Civilization, as we know it, is only about 6,000 years old, and industrialization started in earnest only in the 1800s. While we have accomplished much in this short period of time, it also shows our responsibility as caretakers of the only planet we live on right now. The Ancient Greeks may not have been the oldest civilization, but they are doubtlessly one of the most influential. Among other things, the Greeks came up with the ideas of the ancient Olympics, and developed the concept of democracy and the Senate, and laid the foundations for modern geometry, biology, and physics. Pythagoras, Archimedes, Socrates, Euclid, Plato, Aristotle (Figure 6.9), Alexander the Great, etc. history books are full of these names whose inventions, theories, beliefs, and heroics have had a significant impact on our civilizations.

Figure 6.9. Aristotle, famous philosopher and scientist.

Writing, developed first by people in Sumer, is considered a hallmark of civilization and "appears to accompany the rise of complex administrative bureaucracies or the conquest state". Traders and bureaucrats relied on writing to keep accurate records. Like money, writing was necessitated by the size of the population of a city and the complexity of its commerce among people who are not all personally acquainted with each other. Traditionally, polities that managed to achieve notable military, ideological and economic power defined themselves as "civilized" as opposed to other societies or human groupings outside their sphere of influence, calling them barbarians, savages, and primitives.

The complex culture associated with civilization has a tendency to spread to and influence other cultures, sometimes assimilating them into the civilization. Civilizations can be seen as networks of cities that emerge from pre-urban cultures and are defined by the economic, political, military, diplomatic, social and cultural interactions among them. For example, trade networks were, until the 19th century, much larger than either cultural spheres or political spheres.

7. Civilization, money, energy and time

European civilization was born in Mesopotamia where trade was extensively developed between Asia and Europe. To sustain the exchanges of this trade, they invented writing and money which now is translated in bucks that determine our lives! Money translates or transfers the value we assign to objects or services as a means of exchange for other objects or services we might acquire in a given society. Money has become a universal societal contract in nearly every human society. Money as a token system represents rights to quantities of goods or services. In the original form, the amount of money as cash was limited through the ability of one party to directly make exchange for some amount of the commodity held under promise.

Among the things required in our daily life, some important ones refer to energy: fuel for the car or electricity to light the home, for instance. The corresponding energy is evaluated and translated into a consumed quantity and then, into an amount of money we are able to pay. This is the way things stand now. In here, we are not talking about the theory of relativity and the quantum theory.

Energy is a virtual concept related to physical entities; this is shown in a physical or chemical transformation lasting a certain time as a flow of fluid. The intensity also with the duration of that flow determines the amount of energy delivered. That means that the time elapsed is the key parameter for the considered energy. Otherwise, the stored energy is considered "potential" as long it is not delivered. In such case of latent possibility, time is suspended. Energy is the capacity of a body to do work and nothing more, but money is an expression of energy and we use it to support our life. Time, energy and money are intimately related; time is money, some say.

In our civilizations, as in the previous ones, the basic form of money remains as the metal gold (the dollar being a substitute for it). This is the ultimate refuge currency often used as a reference of state wealth and kept in well-guarded safes. Gold is considered a reference of potential energy that could be converted in any form depending on the needs and the moment.

Gold ingots are rather convenient to represent standard of value, and a symbol of wealth whatever the civilization considered. They are perfectly stainless, high purity, dense and correspond to a high value in a limited volume; they also cannot be destroyed easily. Of course, there are other expensive minerals found in the nature or obtained by synthesis, but gold is the most renowned!

Gold as a metal (Figure 6.10) is largely used in jewelry or more largely in industry, especially electronics where electrical leads and connectors of high quality are required. The electrical conductivity of gold is exceptionally weak (below copper) thus allowing electrical energy transfer very easy. It is largely used in vacuum evaporated layers in the Integrated Circuits as well as printed circuits.

Figure 6.10.   Gold ingots.

The possible production of gold from a more common element, such as lead, has long been a subject of human inquiry, and the ancient and medieval discipline of alchemy often focused on it; however, the transmutation of chemical elements did not become possible until we understood the power of nuclear physics in the 20th century. The first synthesis of gold was conducted by Japanese physicist Hantaro Nagaoka, who synthesized gold from mercury in 1924 by neutron bombardment.

Gold represents a symbol of power, in some way as well as a reference of a potential energy. Gold has a rather high density of 19.3 g/cm^3 and it is thought to have been produced in supernova nucleosynthesis, and from the collision of neutron stars, and to have been present in the dust from which the solar system formed.

Globally, our universe is gradually moving from a state of energy concentration (where some regions of our universe contain more energy than the others) to a state of energy distribution (where all energy in the universe would be equally distributed). The evenness of energy distribution is known as entropy. So, our universe is gradually moving from a state of lower entropy to a state of higher entropy. Closer to home, we also might have to worry about our energy running out, sooner or later.

Chapter 7

Would Energy be Lasting?

Animal species and specifically humans, just as our mobile phones and laptops need to be powered by batteries or some form of energy, require certain things to sustain life and survive. Specifically, we are talking about oxygen and food. They also need to cook food and heat homes during winter. Thermal energy is produced from wood, coal, petroleum, oil or gas which are all consumables that cannot keep up with the demand of the ever-growing world population.

1. Might traditional sources soon be exhausted?

Coal is, by definition, a combustible sedimentary organic rock mainly composed of carbon, hydrogen and oxygen that meets 27% of the global energy needs. The major consumption comes from China (54%), India (12%), and the USA (6%). Coal is the product of a long process of sedimentary carbonization (some 300 million years). Since the 18th century, it has been obtained from mines.

Petroleum (crude oil, or hydrocarbons) obeying similar processes as coal, occurs naturally in geological formations and extracted by drilling. The origin of this fossil fuel comes from zooplankton or algae buried in sedimentary rocks under pressure and heat. It is commonly extracted by drilling and then refined by distillation under different conditions (gasoline, diesel, kerosene). Petroleum is one of the major commonly used energy sources that releases greenhouse gases and consequently leads to the so-called climate change (to be confirmed). Petroleum is well known as a major source of conflicts around the world because of the money in question. All of that argues in favor of the electric car (which looks for the origin of its energy in delocalized sources, mainly nuclear plants).

On top of that, the end product of combustion inevitably manifests as an emission of CO_2 as a result of the combination of carbon atoms with oxygen in the combustion process which results in thermal energy. This is now considered a major environmental pollution even though this carbonic gas is the natural and necessary raw resource for every form of plant life. So, to compensate for this large emission of CO_2 around factories especially, the best solution (free of tax) could certainly be to plant trees instead of burning them! This is a virtuous circle! (Figure 7.1).

The necessary basic resources lie in the ground (coal, gas or fuel) and are extensively exploited. Of course, traditional means such as coal and petroleum or gas (even if largely used) no longer suffice for the rising demand in energy with the ever increasing need to support our lifestyle in World 2.0. On top of that they are accused of causing a global warming of the atmosphere which could generate detrimental climate changes.

To date, it is difficult to distinguish between human-driven climate change and the natural fluctuations of the climate. Are we not a bit presumptuous to assume that we could change the global order of Nature?

In any case, to understand the situation it is necessary to study other sources of energy on as large a scale as possible.

By now, the future of these natural reserves may amount at some 1,187 billion tons of oil equivalent of petroleum. That is to say, only 50 years at the current pace (petroleum and gas). This augurs severe difficulties in the near future.

Concerning ^{235}U consumption, it is expected that 130 years of continuous use could be guaranteed. This delay could be doubled (or more

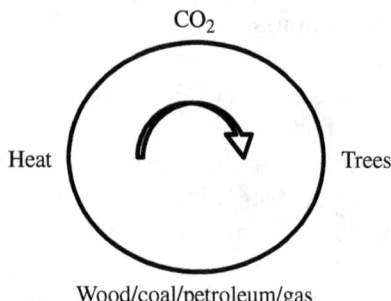

Figure 7.1.   The virtuous circle.

with the fast neutrons technology) through a reprocessing of uranium and plutonium.

Another source of nuclear energy available can also be found in thorium atoms on a rather large scale. All of this means that nuclear energy constitutes a huge and reliable source in the long term, provided that humanity is not afraid of it.

2. Renewable energies

Some sources which are available directly in our natural paradigm have been developed for centuries or even millennia and can be qualified as sustainable (even if not permanent), the first and main one obviously derives from the Sun which indirectly generates winds, rains and light; more indirectly there are renewable energies that can be found from the planet itself (geothermic) and its satellite, the Moon (tidal), the problem being the quantities required and their availability. None of them (even kept together) has the capacity to cover the global need at any one time.

2.1. *Wind*

Wind is known and has been used for millennia (Figure 7.2) to provide a mechanical energy able to drive a boat or power a windmill to produce

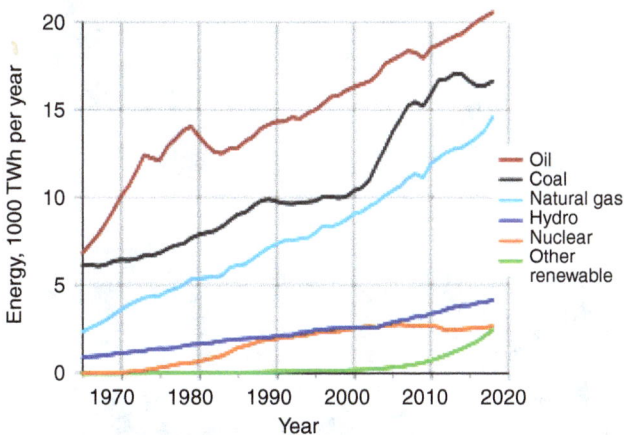

Figure 7.2. World energy consumption over time.

flour or olive oil. In modern times, it appeared that electricity was an ideal intermediate energy, so much so that, windmills were used to move, not a millstone, but a dynamo or an alternator to generate power. This was used for years in the countryside to pump water up from the groundwater.

Then it appeared that modern gigantic (150 m high) wind turbines were to emerge from the 2.0 technology (Figure 7.3).

Many of them have blossomed in the windiest places, even offshore so as to avoid the noise pollution generated. It soon appeared that such a solution was expensive, difficult to maintain and has a much shorter life-time than expected; in addition, their disassembly appeared also very expensive and far from easy to do. This solution was very trendy at the beginning, when people thought it would be a source of free energy. But it is now on the way to being discontinued as long as this expectedly intermittent energy source does not rise to the challenge. So it appears that the free energy we dreamt of has gone with the wind!

It must also be added that, when a windfarm is competing with the inescapable nuclear reactor, it should be appreciated that the latter does not have to lower its stable production to make room for the occasional challenger. A reactor is conceived to deliver a continuous stable power.

Figure 7.3.   Offshore windfarm.

2.2. *Solar*

Solar energy is by far the most abundant source of energy on Earth, but it is so dispersed all over the globe that it requires large surface converters to collect only a little of it.

In a first attempt, mirrors were used to concentrate light and heat but this was not enough in spite of the progress achieved.

When solid state solar cells were obtained with a convenient conversion yield and a limited unit cost, it was preferred to assemble solar farms over fields with very large surfaces (Figure 7.4). However, the conversion yield per unit surface remains relatively limited despite the striking improvements achieved in research and development. Here too, the production of electricity is in no way permanent even in sunny countries and the collected energy requires a huge buffer such as batteries or a reverse pumping mechanism in a dam.

A solution could have been to set such a station on a satellite placed on an orbit where the sun shines day and night; but, what the hell, how are we to transfer the collected energy down to Earth?

Figure 7.4. Solar farms cover very large surfaces.

Latest news: my smartphone just informed me about the project Scope M developed by the US Navy: a beam of 10-GHz microwave was implemented to transfer energy in a wireless process. For now, the yield reached only 60%, but improvements are on the way. This process could be very convenient to transfer energy down (or up to) from a space station (provided that the system aims well!).

2.3. *Water*

Water energy has long been exploited from the ancient water mills along the rivers and by collecting rainfall in huge dams and waterfalls. That works rather well and such participation in the balance is nowadays quite notable. However, the final production of electricity remains inevitably limited to the quantity of water which is obtainable. The main advantage of such a solution is that the available energy may remain stored until being converted. Coupling with solar farms (when achievable) could provide a permanent stable power supply, alternating between solar and water. In France, the large amount of hydropower makes it the third source of available energy after nuclear and thermal.

Another energy coming from water results from the tides which are regularly driven by the gravitational force of the Moon. Here too, the energy is not easily captured on a large scale and remains intermittent. However, the functioning of the turbines may be double: when the tide rises and, afterwards, when the tide lowers. These are a list of options available: you pay your money and pick your choices.

2.4. *Hydrogen*

A promising new development to amortize the fluctuations of the intermittent renewable sources of energy (wind or solar as well) could be to turn them towards the production of "green Hydrogen". This means that these sources of energy could be coupled to a plant for water electrolysis; this is perfectly clean and the resulting gas can be stored before being used as a fuel without any CO_2 emitted.

Hydrogen is the most abundant element in the universe even though there is no natural source of hydrogen on Earth. But it is hard to find in a free state. It must be extracted from other sources such as water, coal, biomass or natural gas. The variety of options for its production and collection, together with its high efficiency in fuel cells and storage capacity over long periods of time, make it a highly valued resource in all sectors of society.

Hydrogen derived from methane — usually from natural gas, but also coal and biomass — was pioneered in World War II by Germany, which has no petroleum deposits. But CO_2 is emitted when manufacturing hydrogen from methane, so it is not climate friendly, and hydrogen manufactured this way is known as gray hydrogen.

The Middle East, which has the world's cheapest wind and solar power, is also angling to be a major player in green hydrogen. "Saudi Arabia has ridiculously low-cost renewable power", said Thomas Koch Blank, leader of the Breakthrough Technology Program at the Rocky Mountain Institute. "The sun is shining pretty reliably every day and the wind is blowing pretty reliably every night. It's hard to beat".[1]

By 2030, some people expect that hydrogen would be capable of powering between 10 and 15 million passenger cars and half a million trucks. Hydrogen fuel cells in electric vehicles are an alternative to battery-powered electric vehicles. They offer greater range, faster recharging times and therefore allow recurring use of the vehicle, something in which current batteries are more limited. In addition to its use in light vehicles, last-mile and heavy road transport, hydrogen can also be used as fuel for rail and sea transport, two sectors where electrification is not technologically feasible at present.

As a form of powerful fuel, Hydrogen must be carefully handled, to prevent leakages and explosion. We all remember the burning of the Graf Zeppelin airship Hindenburg and corresponding casualties. Thereafter it was preferred to use helium (He) instead of hydrogen to fill the balloons; but helium, even if plentiful in the universe, is still rare on earth (mines) and so it is considered to be expensive!

"Green" hydrogen has significantly lower carbon emissions than "gray" hydrogen, which is produced by steam reforming of natural gas and represents 95% of the market. On the other hand, green hydrogen, specifically, that is produced by electrolysis of water, represents less than 0.1% of total hydrogen production. Hydrogen offers the greatest potential to decarbonize difficult-to-abate sectors such as steel, cement

---

[1] *The new fuel to come from Saudi Arabia.* BBC - Homepage (2020). Retrieved December 12, 2022 from https://www.bbc.com/future/article/20201112-the-green-hydrogen-revolution-in-renewable-energy.

and heavy-duty transport. Green hydrogen has been used in transportation, heating, and in the natural gas industry, and can be used to produce green ammonia required in the production of fertilizers. Additionally, hydrogen-powered aircrafts are already being designed by Airbus, with a planned release of the first commercial aircraft by 2035.

As a matter of fact, storing hydrogen as a gas would require huge storage tanks so it becomes better to have it compressed or even cooled down to the liquid phase which can be reached at some 4 K (i.e., −269°C) under normal pressure.

Among others, a consortium of companies has announced a $30 billion project in Oman which would become one of the largest hydrogen facilities in the world. Construction will begin in 2028, and by 2038, the project will be powered by 25 GW of wind and solar energy.

Spain also is in a privileged position for large-scale production, due to its capacity to generate renewable power. It also has an infrastructure network already capable of transmission and storage, which will mean that energy transition can take place at the lowest possible cost, a key factor for ensuring that this process is fair and inclusive, and guarantees the supply and flexibility of the energy system.

3. Nuclear energy

So due to the lack of free, abundant supply of clean and renewable energy, it becomes essential to switch to other less virtuous but more generous and adapted solutions such as nuclear power. These methods dive in the very core of the atom and allow very large (some say unlimited) production of energy provided that some adequate precautions are taken. No production of CO_2 results, but radioactive wastes have to be buried in special facilities either at ground level or deeper in underground repositories for the long-life ones. The amount of free energy contained in nuclear fuel is millions of times the amount of free energy contained in a similar mass of chemical fuel such as gasoline, making nuclear fission a very dense source of energy.

At present, 453 reactors are in operation throughout the world, mainly under the classification PWR or BWR.[2] Globally they develop some 400,000 MW, about 10.5% of the world's consumption of electricity.

---

[2]Pressured water reactor; boiling water reactor.

Most of this production takes place in the US, France, China, Japan and Russia. The largest concentration of nuclear electricity is in France.

In spite of some dire predictions, the number of major nuclear accidents, on a global scale, remains rather limited to date:

- Three Mile Island (USA, 1979); no casualties.
- Chernobyl (USSR, 1986); massive radioactive pollution and some severe contamination; no direct casualties. Causes were found to be very bad technical maintenance.
- Fukushima (Japan, 2011); consequence of a tsunami following an earthquake; no direct casualties, only some severe radioactive contamination.

The impact of these disasters was so large in the public opinion that a kind of phobia attributed to nuclear energy spread over some nations (such as Germany for instance) to the point that political U-turns were taken on long-standing national policy on power supply.

As a matter of fact, generally speaking, the more powerful a source of energy (whatever it may come from), the more dangerous it could be; it is the fundamental rule in all circumstances.

Many nuclear reactors are also involved in the naval propulsion, essentially in military applications of submarine or aircraft carriers powered through vapor turbines. However, no problems have been reported to date.

Nonetheless, the most promising possibility could arise from the phenomenon of the inertial fusion of light atoms (deuterium and tritium) as it appears in the stars such as the Sun. A very instable plasma must be initiated and maintained in order for the reaction to develop. This was already achieved explosively in a H-bomb but not mastered, to date, in a civilian application.

Two different solutions are being explored at present: the first one uses what is called "inertial confinement", the other one uses a laser-focused impact on micro balls.

The main advantage of this source of energy (aside from the expected delivered power) lies in the resulting almost total absence of radioactive waste.

A technical operational solution has not yet been found in spite of great efforts and timelines such as the ongoing international program ITER in Cadarache, France. It seems, also to date, that this solution (when reached) could be the sole way to meet the needs of an unlimited energy demand all over the world.

Though ITER, itself, is not intended to generate electricity, its successors, i.e., DEMO reactors would generate electricity latest by the 2050s. Many government and private initiatives have started to achieve the same target faster. Chinese Fusion Engineering Test Reactor (CFETR) uses ITER research and will be commercially operational by 2050 though a prototype will be ready by 2035. K-DEMO of Korea will also use ITER data and will start electricity generation in 2050s.

Now, a record of 59 megajoules (MJ) has been reached at 5 s with a stable plasma, in the JET[3] tokamak at Oxford, which entailed a temperature ten times higher than in the core of the Sun.

4. Energy for space applications

While dealing with satellites and other space developments, energy is the key enabler that powers everything that moves. The most obvious question remains the mode of propulsion. Usually, rockets use chemical energy, burning solid propellants or, more likely, liquid rocket fuels in a nozzle and thus give a mechanical recoil reaction. This push is required to help the rocket cross the atmosphere and move in the cosmic vacuum where they obey only the mechanical laws of inertia. Any change in this movement requires adding energy and, when coming back on Earth, all this accumulated energy must be spent in friction (then heat) with the atmosphere.

During a space mission, the space vehicle has to carry sufficient energy to make things happen. Solar cells make a significant contribution to slowing down the need to carry the heavy payload of ever more batteries. On the other hand, the development of small nuclear reactors has, to a certain extent, helped to power some long-range one-way missions. In the case of a settlement on the Moon (or other places), this would take the form of a complete and sustainable installation of an energy plant. Since nuclear plants can be used not only as a source of electric power, but also as a source of heat and propulsion, they might be our only choice to reach bases beyond planet Earth and support life forms and productive activities there.

---

[3] Joint European Torus.

However, due to the intensity of sunlight decreasing with the square of the distance from the Sun, solar power becomes too weak beyond a certain distance to be the source of power for the spacecraft once it is propelled into the outer solar system and beyond; it will need a different source of energy to power and sustain the systems. (The Juno spacecraft, launched in 2011, will be the first mission to observe Jupiter using solar power; however, it will need to employ three huge solar panels of 2.7 m × 8.9 m size to meet its power requirements.) Also, solar power will be a challenge for landers and rovers that must operate on rocky surfaces or in the dark side of planets in long-duration shadows.

Nuclear power systems have been in use for more than five decades, enabling us to gain sufficient knowhow for deployment in distant planets and interstellar space. The two basic types of nuclear power supply used in space are "nuclear reactors" and "radioisotope sources".

Rather, this technology relies on thermoelectric materials (special kinds of semiconductors to generate electric current when one end is kept hotter than the other end). The greater the temperature difference, the stronger electricity is produced. These systems, generally referred to as Radioisotope Thermoelectric Generators (RTGs), are known to be simple and very reliable since they do not involve any moving parts. They are also long-lasting power sources since they operate as long as the isotope produces a useful level of heat.

Plutonium-238, with a half-life of 88 years, is the most widely used radioisotope in RTGs. With its high heat capacity, it allows heat sources to be compact while resulting in useful levels of energy to provide the heat-to-electricity conversion process. One kilogram of ^{238}Pu can generate approximately 560 W of thermal power.

The first nuclear-powered spacecraft, the navigational satellite Transit 4A, was launched more than 50 years ago in 1961. Since then, dozens of spacecraft have been launched carrying RTGs and have explored the Moon and almost every planet in the solar system. In many of these cases, the radioisotope was used not only for power generation, but also to keep the spacecraft warm in frigid environments on planetary surfaces or in deep space, which is quite critical for efficient operations.

A good deal of effort has been invested to make today's RTGs more or less impervious to the most likely launch accident scenarios. One important accident involving RTGs was the heroic Apollo 13 mission, which is known as the "successful failure" in space history. This mission fortunately ended with safe return of the astronauts to Earth, however not only they but also their RTG, which was intended to be left behind on the Moon's surface, made it back to Earth. The RTG and its 3.9 kg of pluto-nium dioxide plunged into the Tonga Trench in the Pacific Ocean, where it will remain radioactive for the next 2,000 years. Subsequent water test-ing has shown the RTG is not leaking radioactivity into the ocean.

Satellites in space can assist with oil and gas prospecting while also ensuring renewable energy infrastructure runs more effectively, through wind field monitoring and "sunshine maps". Space weather forecasts are also relevant, safeguarding energy infrastructure such as power grids from harmful power surges, or oil pipelines from current-driven corrosion.

5. Energy for data centers

It clearly appears to us that the World 2.0 that we are building is driven by data, like blood in our body; and as data centers grow in size, capacity and complexity, they become hungrier and hungrier for energy. Huge strides in energy efficiency including a shift to efficient "hyperscale" data centers have helped to limit data center electricity demand growth globally. At the local level, however, these large hyperscale data centers represent huge electricity demand loads, adding pressure to electricity grids and increas-ing the challenge of energy transitions. Some media headlines warn that a "'Tsunami of data' could consume one-fifth of global electricity by 2025".[4]

Contrary to these alarming headlines, data centers worldwide only consumed around 200 terawatt-hours (TWh) in 2018, or about 1% of global electricity use. Their energy use has been flat since 2015, while global internet traffic tripled. Increasingly efficient IT hardware and a major shift to hyperscale data centers have helped to keep electricity demand flat, despite exponential growth in demand for data center ser-vices. A hyperscale data center can require 100–150 MW of grid capacity and consume hundreds of GWh of electricity annually.

Finally, waste heat from data centers could help to heat nearby com-mercial and residential buildings, reducing their energy use from other

---

[4]"Tsunami of data" could consume one fifth of global electricity by 2025. | Environment. *The Guardian*. (2017). Retrieved December 13, 2022 from https://www.theguardian.com/environment/2017/dec/11/tsunami-of-data-could-consume-fifth-global-electricity-by-2025.

sources. Strong efforts in energy efficiency as well as initiatives on demand response and waste control in heat utilization can help minimize the impacts of large data centers on the grid as well as the environment.

In 2018, Google (10 TWh) and Apple (1.3 TWh) purchased or generated enough renewable electricity to match 100% of their data center energy consumption. Facebook data centers consumed 3.2 TWh in 2018, of which 75% came from renewables. Amazon and Microsoft sourced about half of the data center electricity from renewables.

As the number of global internet users has grown, so too has demand for data center services, giving rise to concerns about the growing demand to sustain the quest for energy from data centers. Between 2010 and 2018, global Internet Protocol (IP) traffic — the quantity of data traversing the internet — increased more than ten-fold, while global data center storage capacity increased by a factor of 25 in parallel. This to say that said, in the coming decade, a significant risk exists that rapidly growth of information services and computational-intensive applications like AI in particular, will outpace the efficiency gains that have up-to-now been able to keep pace with the exponential growth of the data industry.

6. Biological energy

Bioenergy is a renewable energy created from natural, biological sources. Many natural sources, such as plants, animals, and their byproducts, could be potential resources of renewable energy (Figure 7.5). Modern technology even makes landfills or waste zones potential bioenergy resources.

The very origin of all these renewable energy sources could be traced back to the Sun that distributed radiated energy transformed through the process of photosynthesis that convert heat and photons and stored into carbohydrate molecules such as sugars and starches, and can be later restored. That means a virtuous circle operates.

Plants absorb light primarily using the pigment chlorophyll. The green part of the light spectrum is not absorbed but is reflected, which is why most plants are green. Besides chlorophyll, plants also use pigments such as carotenes and xanthophylls. Algae also use chlorophyll, but various other pigments are also possible, such as phycocyanin, carotenes, and xanthophylls in green algae, phycoerythrin in red algae (rhodophytes) and fucoxanthin in brown algae; and diatoms similarly exist in a wide variety of colors.

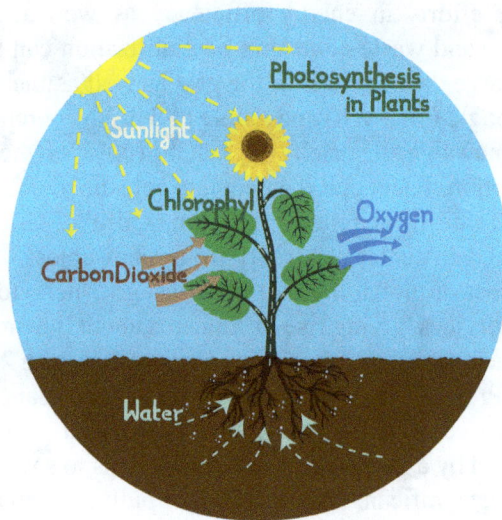

Figure 7.5.   Scheme of photosynthesis.

Biomass is one type of renewable resource that can be converted into liquid fuels — known as biofuels — for transportation. Biofuels include cellulosic ethanol, biodiesel, and renewable hydrocarbon "drop-in" fuels. The two most common types of biofuels in use today are ethanol and biodiesel. Biofuels can be used in airplanes and most vehicles that are on the road. These renewable transportation fuels are functionally equivalent to petroleum fuels but lower in carbon intensity for use in our vehicles and airplanes.

Producing advanced biofuels (e.g., cellulosic ethanol and renewable hydrocarbon fuels) typically involves quite a complex multistep process. First, the tough rigid structure of the plant cell wall — which includes the biological molecules cellulose, hemicellulose, and lignin bound tightly together — must be broken down. This can be accomplished in one of two ways: high temperature deconstruction (pyrolysis, gasification, hydrothermal liquefaction) or low temperature deconstruction (use of biological catalysts called enzymes or chemicals which break down feedstocks into intermediates).

7. Thermal energy and oxygen

When dealing with thermal energy, usually people think of wood, coal, fuel, and forget about the essential counterpart: oxygen. Unlike hydrogen, oxygen is the Earth's most abundant element as a diatomic molecule; some 21% of the atmosphere thus contributing to essential energy released in combustion and aerobic cellular respiration; many major classes of organic molecules in living organisms contain oxygen atoms.

Most of the mass of living organisms consists of oxygen as a component of water, the major constituent of lifeforms. Oxygen is continuously replenished in Earth's atmosphere by photosynthesis, which make uses of sunlight as source of energy to produce oxygen from water and carbon dioxide. Lavoisier conducted the first adequate quantitative experiments on oxidation and provide us with the first correct explanation of how combustion works. Oxygen is also a byproduct of the electrolysis of water, giving rise to hydrogen in the process.

The main driving factor of the oxygen cycle is photosynthesis, which is responsible for modern Earth's atmosphere. Photosynthesis disassembles the CO_2 molecule and releases oxygen into the atmosphere, while respiration, decay, and combustion remove it from the atmosphere. In the present equilibrium, production and consumption occur at the same rate (which is recommended).

This cycle, along with many other geological cycles, is driven by the energy of the Sun. But the Sun has not been burning forever; and its Energy must also have come from a prior source, which is a topic that merits further study.

Chapter 8

Where Does Energy Come From?

After much hesitation, the first thought that came into our mind was that we could not finish this book without examining the origin of Energy. However, finally we decided to focus on our mind on this highly controversial topic. Energy is at the root of everything in our world, but where did it come from at the very beginning? From the teachings of religions, there is no firm scientific conviction, but only possible theories about the starting point (if any) and the starting time. Known as the "Big Bang", in default of knowing what could have preexisted and who pushed the red button.

However, it is difficult to talk about the Big Bang without touching on cosmology, relativity, spacetime warping or other complex theories on black hole behavior.

We are all based on the principle of causality, but is it really essential? In the current circumstances, is there an acceptable proposition about the origin of the world. Is a cause needed to start the story? Would such hypotheses be acceptable? Would our universe be unique or be accompanied by other universes? The questions are endless but nevertheless in some way connect to the concept of energy in the background.

1. No answer, but theories

Everything we can elucubrate stems from the time $(0 + \varepsilon)$; then we have to turn the clock back, starting from the elements we find in the present era, on Earth as well as in faraway space.

There are several hypotheses about the origins and/or cyclic nature of the universe, but none has gained preeminence. Some options are Nothing, There is no direct relationship but they both relate to the question of how order can be created out of chaos. Fundamentally, to build a universe requires three things: space, mass and energy.

The physics community has debated the various multiverse theories for a long time. Prominent physicists are divided about whether any other universes exist outside of our own. Some physicists say the multiverse is not a legitimate topic of scientific inquiry. Concerns have been raised about whether attempts to exempt the multiverse from experimental verification could erode public confidence in science and ultimately damage the study of fundamental physics. Some have argued that the multiverse is a philosophical notion rather than a scientific hypothesis because it cannot be empirically falsified. The ability to disprove a theory by means of scientific experiment is a critical criterion of the accepted scientific method. https://en.wikipedia.org/wiki/Multiverse-cite_note-Richard_P._Feynman_on_The_Scientific_Method-7 Paul Steinhardt has famously argued that no experiment can rule out a theory if the theory provides for all possible outcomes.

In 2007, Nobel laureate Steven Weinberg suggested that if the multiverse existed, "the hope of finding a rational explanation for the precise values of quark masses and other constants of the standard model that we observe in our Big Bang is doomed, for their values would be an accident of the particular part of the multiverse in which we live".[1]

Around 2010, scientists such as Stephen M. Feeney analyzed Wilkinson Microwave Anisotropy Probe (WMAP) data and claimed to find evidence suggesting that this universe collided with other (parallel) universes in the distant past. However, a more thorough analysis of the data from the WMAP and from the Planck satellite, which has a resolution three times higher than WMAP, did not reveal any evidence that was statistically significant to support the hypothesis of bubble universe collision. In addition, there was no evidence of any gravitational pull of other universes on ours.

As Einstein realized, two of the main ingredients needed to make a universe — mass and energy — are basically the same thing. So, where

---

[1] Steven, W. (2013). *Physics: What We Do and Don't Know. The New York Review of Books*. Retrieved December 12, 2022 from https://www.nybooks.com/articles/2013/11/07/physics-what-we-do-and-dont-know/.

did all this energy and space come from? They came spontaneously into being in an event we call the Big Bang. But there are obvious questions here. What happened before the beginning? What was God doing before? Does the universe have a beginning (Kant)?

Previously the generally accepted answer was the oscillating universe theories. The idea was that the Big Bang would be followed by a Big Crunch. The universe would expand for a time and then contract, forming another Big Bang. This allowed for the concept of material realism that matter and energy were basically eternal and self-existent.

Then in 1997 and 1998, studies of supernovae provided evidence that the rate of expansion of the universe was increasing. This empirical data indicated that the contraction of the universe needed for the oscillating universe theories will never exist. The implication of the supernova studies is that the universe is an open system, and there must be something that exists outside of the space, time, matter, energy continuum of our present universe. And the following question: Where is God trying to hide from us?

There are many theories proposed to explain the origin of the universe. The Big Bang simply denotes the instant at which some primordial, currently unknown, type of energy was instantly converted into matter through a large, universal inflation to create the elements and particles that we know to be a part of the standard model. What that primordial energy was or what made it "unstable" is not known. You could look into research done on quantum fluctuations or string theory and parallel universe collisions if you want to try and decipher the current working theories. If you find a way to prove a theory experimentally for the state of the universe before the Big Bang, you would be on your way to collect a Nobel prize, I'm sure :-).

Final Contraction of the Previous Universe, and Chaos. Author Terry Pratchett (RIP) made some interesting speculations. Dr. Stephen Hawking[2] has one of the most balanced reviews of current theories and possibilities. Check out his very readable books on the History of Time. It may, or may not, have had a "start".

Observational evidence of a very dense beginning came in 1965 with the discovery of a faint background of microwaves throughout space. The only reasonable interpretation of that background is that it is radiation left over from an earlier very hot and dense state. As the universe expanded,

---

[2]Hawking, S. (2018). *Brief Answers to the Big Questions*. John Murray.

the radiation would have cooled until it is just the faint remnant we discovered today (1993) with the Cosmic Background Explorer (COBE) satellite.

Another "steady state theory"[3,4] was that new galaxies would form from matter continuously being created throughout space, and the universe would have looked the same at all times.

This would be because it is still unclear on why and how it happened. We have a number of possibilities on how and why there was a Big Bang, but no scientist has ever been able to find a formula or a clear explanation. That is why the Big Bang Theory is not a law.

What exists outside of our universe is beyond scientific observation or experimentation. One theory is the multi-universe theory. A variation of the multi-universe theory is the white hole black hole theory and the idea of worm holes. Chaos theory or Nothing Really isn't Nothing theory has also been proposed. Scientists are reluctant to give up the concepts that matter and energy are all that exists because matter and energy are all that science can investigate. That there is something non-material outside of the universe is regarded as unscientific.

The main evidence consistent with the Big Bang Theory is Cosmic Microwave Background (CMB) radiation. CMB radiation was unknown during the development of the theory, is consistent with Einstein's Theory of Relativity, and was discovered while scientists were looking for something else. Thus, its discovery and agreement with the theoretical implications of both the Big Bang theory and General Relativity make it the single best piece of current evidence that such an event occurred.

According to the theories of physics, if we were to look at the Universe one second after the Big Bang, what we would see is a 10 billion-degree wave of neutrons, protons, electrons, anti-electrons (positrons), photons, and neutrinos. Then, as time went on, we would see the Universe cool, the neutrons either decaying into protons and electrons or combining with protons to make deuterium (an isotope of hydrogen). As it continued to cool, it would eventually reach the temperature where

---

[3] Bondi, H. & Gold, T. (1948). The steady-state theory of the expanding universe. *Monthly Notices of the Royal Astronomical Society, 108*(3), 252–270. https://doi.org/10.1093/mnras/108.3.252.

[4] Hoyle, F. (1948). A new model for the expanding universe. *Monthly Notices of the Royal Astronomical Society, 108*(5), 372–382. https://doi.org/10.1093/mnras/108.5.372.

electrons combined with nuclei to form neutral atoms. Before this "recombination" occurred, the Universe would have been opaque because the free electrons would have caused light (photons) to scatter the way sunlight scatters from the water droplets in clouds. But when the free electrons were absorbed to form neutral atoms, the Universe suddenly became transparent. Those same photons — the afterglow of the Big Bang known as cosmic background radiation — can be observed today.

The National Aeronautics and Space Administration (NASA) has launched two missions to study the cosmic background radiation, taking "baby pictures" of the Universe only 400,000 years after it was born. The first of these was COBE. In 1992, the COBE team announced that they had mapped the primordial hot and cold spots in cosmic background radiation. These spots are related to the gravitational field in the early Universe and form the seeds of the giant clusters of galaxies that stretch hundreds of millions of light years across the Universe. This work earned NASA's Dr. John C. Mather and George F. Smoot of the University of California the 2006 Nobel Prize for Physics.

The second mission to examine the cosmic background radiation was WMAP. With greatly improved resolution as compared to COBE, WMAP surveyed the entire sky, measuring temperature differences of the microwave radiation that is nearly uniformly distributed across the Universe. The picture shows a map of the sky, with hot regions in red and cooler regions in blue. By combining this evidence with theoretical models of the Universe, scientists have concluded that the Universe is "flat", meaning that, on cosmological scales, the geometry of space satisfies the rules of Euclidean geometry (e.g., parallel lines never meet, the ratio of circle circumference to diameter is pi, etc.).

A third mission, Planck, led by the European Space Agency (ESA) with significant participation from NASA, was. launched in 2009. Planck is making the most accurate maps of the microwave background radiation yet. With instruments sensitive to temperature variations of a few millionths of a degree, and mapping the full sky over nine wavelength bands, it measures the fluctuations of the temperature of the CMB with an accuracy set by fundamental astrophysical limits.

But all this leaves unanswered the question of what powered inflation. One difficulty in answering this question is that inflation was over well before recombination, and so the opacity of the Universe before recombination is, in effect, a curtain drawn over those interesting very early events. Fortunately, there is a way to observe the Universe that does not involve photons at all. Gravitational waves, the only known form of information that can reach us undistorted from the instant of the Big Bang, can carry information that would be impossible by other means. Several missions are being considered by NASA and ESA that will look for the gravitational waves from the epoch of inflation.

The Hubble telescope was able to measure galaxies that are moving away; the further they are the faster they move: the universe is expanding. So, they must have been closer together in the past. From the present rate of expansion we can estimate that they must have been close together about 10–15 billion years ago.

Astronomers combine mathematical models with observations to develop workable theories of how the Universe came into being. The mathematical underpinnings of the Big Bang theory include Albert Einstein's general theory of relativity along with standard theories of fundamental particles. Today NASA spacecrafts such as the Hubble Space Telescope and the Spitzer Space Telescope continue to measure and search for the expansion of the Universe. One of the goals has long been to decide whether the Universe will expand forever, or whether it will stop someday, turn around, and collapse in a "Big Crunch?"

The more recent James Webb telescope is the biggest one built so far (6.5 m in diameter); it was launched at the end of 2021, covers a wavelength range 0.6–28 μm, and was placed into an orbit around the Lagrange point L_2 at a distance of 1.5 million km from Earth.

2. What could be observable today?

What we have so far is a number of powerful means for investigation, but what have we uncovered so far about our remote environment?

2.1. *Black holes*

A black hole is a region in space where the pulling force of gravity is so strong that light is unable to escape. The strong gravity occurs because matter has been compressed into a tiny space. This compression can take place at the end of a star's life. Some black holes are a result of dying stars. A black hole is not truly a hole. It is quite the opposite.

These objects have so much mass — and therefore gravity — that nothing can escape them, not even light. That makes them some of the most extreme objects in the universe. Most black holes form after a giant star, one at least 10 times as massive as our sun, runs out of fuel and collapses. The star shrinks and shrinks and shrinks. Eventually, it forms a tiny dark point. This is known as a stellar-mass black hole. Now much smaller than the star that made it, this black hole still has the same mass and gravity.

Black-hole cosmology is a cosmological model in which the observable universe is the interior of a black hole existing as one of possibly many universes inside a larger universe.[5] This includes the theory of white holes, which are on the opposite side of space-time.

As nothing can escape a black hole — not visible light, X-rays, infrared light, microwaves or any other form of radiation — black holes are invisible. As a result, astronomers have had to "observe" them by studying how they affect their surroundings.

Black holes formed by the collapse of individual stars are relatively small, but incredibly dense. One of these objects packs more than three times the mass of the sun into the diameter of a city. This leads to a massive amount of gravitational force pulling on objects around the object. Stellar black holes then consume the dust and gas from their surrounding galaxies, which keeps them growing in size. Supermassive black holes could arise from large clusters of dark matter. This is a substance that we can observe through its gravitational effect on other objects; however, we do not know what dark matter is composed of because it does not emit light and cannot be observed directly.

Albert Einstein first predicted the existence of black holes in 1916, with his General Theory of Relativity. The term "black hole" was coined many years later in 1967 by American astronomer John Wheeler. After decades of black holes being known only as theoretical objects, the first physical black hole ever discovered was spotted in 1971.

2.2. *Dark matter*

Dark matter is a hypothetical form of matter thought to account for approximately 85% of the matter in the universe. Dark matter is called "dark" because it does not appear to interact with the electromagnetic field, which means it does not absorb, reflect, or emit electromagnetic radiation (like light) and is, therefore, difficult to detect. Various

---

[5] https://en.wikipedia.org/wiki/Multiverse-cite_note-67.

astrophysical observations — including gravitational effects which cannot be explained by currently accepted theories of gravity unless more matter is present than can be seen — imply dark matter's presence. For this reason, most experts think that dark matter is abundant in the universe and has had a strong influence on its structure and evolution.

The primary evidence for dark matter comes from calculations showing that many galaxies would behave quite differently if they did not contain a large amount of unseen matter. Some galaxies would not have formed at all and others would not move as they currently do. Other lines of evidence include observations in gravitational lensing (see footnote 4) and the CMB radiation, along with astronomical observations of the observable universe's current structure, the formation and evolution of galaxies, mass location during galactic collisions, (see footnote 5) and the motion of galaxies within galaxy clusters. In the standard Lambda-CDM model of cosmology,

Because no one has directly observed dark matter yet — assuming it exists — it must barely interact with ordinary baryonic matter and radiation except through gravity. Most dark matter is thought to be non-baryonic; it may be composed of some as-yet-undiscovered subatomic particles.[6] The primary candidate for dark matter is some new kind of elementary particle that has not yet been discovered, particularly weakly interacting massive particles (WIMPs). Though axions have drawn renewed attention due to the non-detection of WIMPs in experiments. Many experiments to directly detect and study dark matter particles are being actively undertaken, but none have yet succeeded. Dark matter is classified as "cold", "warm", or "hot" according to its velocity (more precisely, its free streaming length). Current models favor a cold dark matter scenario, in which structures emerge by the gradual accumulation of particles.

Although both dark matter and ordinary matter are matter, they do not behave in the same way. In particular, in the early universe, ordinary matter was ionized and interacted strongly with radiation via Thomson scattering. Dark matter does not interact directly with radiation, but it does affect the CMB radiation by its gravitational potential (mainly on large scales) and by its effects on the density and velocity of ordinary matter. Ordinary and dark matter perturbations, therefore, evolve differently with time and leave different imprints on the CMB radiation.

---

[6]https://en.wikipedia.org/wiki/Dark_matter-cite_note-15.

A stream of observations in the 1980s supported the presence of dark matter, including gravitational lensing of background objects by galaxy clusters, the temperature distribution of hot gas in galaxies and clusters, and the pattern of anisotropies in the CMB. According to consensus among cosmologists, dark matter is composed primarily of a not yet characterized type of subatomic particle. The search for this particle, by a variety of means, is one of the major efforts in particle physics.

Unlike normal matter, dark matter does not interact with the electromagnetic force. This means it does not absorb, reflect or emit light, making it extremely hard to spot. In fact, researchers have been able to infer the existence of dark matter only from the gravitational effect it seems to have on visible matter.

Structure formation refers to the period after the Big Bang when density perturbations collapsed to form stars, galaxies, and clusters. Prior to structure formation, the Friedmann solutions to general relativity describe a homogeneous universe. Later, small anisotropies gradually grew and condensed the homogeneous universe into stars, galaxies and larger structures. Ordinary matter is affected by radiation, which is the dominant element of the universe at very early times.

The hypothesis of dark matter has an elaborate history. In a talk given in 1884, Lord Kelvin estimated the number of dark bodies in the Milky Way from the observed velocity dispersion of the stars orbiting around the center of the galaxy. By using these measurements, he estimated the mass of the galaxy, which he determined is different from the mass of visible stars. Lord Kelvin thus concluded "many of our stars, perhaps a great majority of them, may be dark bodies".

2.3. *Dark energy*

Dark energy makes up approximately 68% of the universe and appears to be associated with the vacuum of space. It is distributed evenly throughout the universe, not only in space but also in time — in other words, its effect is not diluted as the universe expands. The even distribution means that dark energy does not have any local gravitational effects, but rather a global effect on the universe as a whole. This leads to a repulsive force, which tends to accelerate the expansion of the universe. The rate of expansion and its acceleration can be measured by observations based on Hubble's Law. These measurements, together with other scientific data, have confirmed the existence of dark energy and provide an estimate of just how much of this mysterious substance exists.

Mass creates gravity, gravity creates pull, and the pulling must slow the expansion. But supernovae observations showed that the expansion of the Universe, rather than slowing, is accelerating. Something, not like matter and not like ordinary energy, is pushing the galaxies apart. This "stuff" has been dubbed dark energy, but to give it a name is not to understand it. Whether dark energy is a type of dynamical fluid, hitherto unknown to physics, or whether it is a property of the vacuum of empty space, or whether it is some modification to general relativity is not yet known.

The Dark Energy Survey (DES) is an astronomical survey designed to constrain the properties of dark energy. It uses images taken in the near-ultraviolet, visible, and near-infrared to measure the expansion of the Universe using Type Ia supernovae, baryon acoustic oscillations, the number of galaxy clusters, and weak gravitational lensing.

Thus, we know that general relativity is necessarily incomplete. It breaks down at the conditions of the early universe, so we currently have no physical model to explain that time. Rather, we know that the early universe expanded, and the Big Bang is the time that perplexes cosmologists. Some theories, such as quantum gravity, have emerged in an effort to explain the Big Bang; however, we currently have little understanding of what it actually was.

In physical cosmology and astronomy, dark energy is an unknown form of energy that affects the universe on the largest scales. The first observational evidence for its existence came from measurements of supernovas, which showed that the universe does not expand at a constant rate; rather, the universe's expansion is accelerating. Understanding the universe's evolution requires knowledge of its starting conditions and composition. Before these observations, scientists thought that all forms of matter and energy in the universe would only cause the expansion to slow down over time. Measurements of the CMB radiation suggest the universe began in a hot Big Bang, from which general relativity explains its evolution and the subsequent large-scale motion.

3. Universe limits explorations

The first telescope is said to have been invented in 1608 in the Netherlands by an eyeglass-maker named Hans Lippershey. The Orbiting Astronomical Observatory 2 was the first space telescope launched on December 7,

1968. As of February 2, 2019, there were 3,891 confirmed exoplanets discovered. The Milky Way is estimated to contain 100–400 billion stars and more than 100 billion planets. There are at least 2 trillion galaxies in the observable universe. HD1 is the most distant known object from Earth, reported as 33.4 billion light-years away.

The research conducted by national space exploration agencies, such as NASA and Roscosmos, is one of the reasons supporters cite to justify government expenses. Economic analyses of the NASA programs often showed ongoing economic benefits (such as NASA spin-offs), generating many times the revenue of the cost of the program. It is also argued that space exploration would lead to the extraction of resources on other planets and especially asteroids, which contain billions of dollars that worth of minerals and metals. Such expeditions could generate a lot of revenue. In addition, it has been argued that space exploration programs help inspire youth to study science and engineering. Space exploration also gives scientists the ability to perform experiments in other settings and expand humanity's knowledge.[7]

The word *observable* in this sense does not refer to the capability of modern technology to detect light or other information from an object, or whether there is anything to be detected. It refers to the physical limit created by the speed of light itself. No signal can travel faster than light, hence there is a maximum distance (called the particle horizon) beyond which nothing can be detected, as the signals could not have reached us yet. Sometimes astrophysicists distinguish between the *visible* universe, which includes only signals emitted since recombination (when hydrogen atoms were formed from protons and electrons and photons were emitted) — and the *observable* universe, which includes signals since the beginning of the cosmological expansion (the Big Bang in traditional physical cosmology, the end of the inflationary epoch in modern cosmology).

The limit of observability in our universe is set by a set of cosmological horizons which limit — based on various physical constraints — the extent to which we can obtain information about various events in the universe. The most famous horizon is the particle horizon which sets a limit on the precise distance that can be seen due to the finite age of the universe. Additional horizons are associated with the possible future extent of observations (larger than the particle horizon owing to the

---

[7] https://en.wikipedia.org/wiki/Space_exploration-cite_note-74.

expansion of space), an "optical horizon" at the surface of last scattering, and associated horizons with the surface of last scattering for neutrinos and gravitational waves.

Telescopes and antennas around the world constantly scan the night sky, observing across the electromagnetic spectrum as part of ground-breaking research into the nature of our universe. While the first tele-scopes relied solely on visible light, now observatories can detect radio and microwaves. An area that has spurred intense research is the CMB radiation, a remnant of the Big Bang. Highly sensitive telescopes and antennas located far from sources of electromagnetic interference (EMI) observe deep space at this spectrum to research the creation, expansion and composition of our universe.

As the Earth rotates, telescopes and antennas must move axially and elevate to keep aligned with a particular observed section of the sky. In addi-tion, in some cases the dishes and receivers must be rotated to ensure accu-rate reception of signals from space. The rotating ranges of the telescopes are limited because of cabling, cooling hoses and mechanical constraints.

The sharpest eye that astronomers have on the universe today is the European Southern Observatory's Very Large Telescope (VLT), a €500 million array of four 8-m diameter telescopes that sit on the 2,500-m high peak of Cerro Paranal. These telescopes have produced the first direct images of planets outside our solar system, weighed distant stars and made important observations of black holes.

But the VLT cannot see small planets. The faintest objects at the edge of the universe, which give astronomers clues about how the universe began, are barely detectable. Bigger telescopes can gather more light and, therefore, produce much more information. The VLT will be able to image faint, previously unseen objects such as cold stars, baby galaxies and small planets, while collecting far more detail on the more familiar ones. The VLT will also be able to take pictures of individual stars in faraway galaxies for the first time. "All galaxies are composed of stars and we understand stars very well so we can use them to understand galaxies. That's interesting if you want to know how the first galaxies developed", said Dr. Andreas Kaufer, director of the European Southern Observatory's (ESO) telescopes at Cerro Paranal.[8]

---

[8]*New telescope could reveal secrets of the universe. Mail & Guardian* — Africa's Best Read. (2006). Retrieved December 12, 2022 from https://mg.co.za/article/2006-08-05-new-telescope-could-reveal-secrets-of-the-universe/.

Objects in the universe emit other electromagnetic radiation such as infrared, X-rays and Gamma rays. These are all blocked by the Earth's atmosphere, but can be detected by telescopes placed in orbit round the Earth. Telescopes in space can observe the whole sky and operate both by day and by night. However, they are difficult and expensive to launch and maintain, let alone if anything went haywire; we would need to fly astronauts to fix them! Once again, energy constraint rears its ugly head.

Conclusion

Would There Be a Conclusion?

Energy is a virtual concept that accompanies physical reality. It comes from everywhere to give meaning and a quantitative explanation for any phenomenon in our world. Closely related to another virtual concept which we name "force", this enables us to give a formal description of what is happening and, in a way, what could happen.

Energy can fit any phenomenon from biological to nuclear, fire to heat, as well as space mechanics. Energy is at the heart of human evolution from implementation of tools, machines, to complex systems. We are, thus far, absolutely dependent on energy.

Energy needs to be understood and is accompanied in its implementation by improvements in our scientific knowledge. Some say that energy is just a gift of God at the root of everything from the original Big Bang which gave rise to our world. Every civilization has been stimulated by some kind of new source of energy to the point that our present world is totally dependent on it.

The very beginnings of our present comfort emerged from the energy stored in a stone, the flint, that gave rise to fire. Then tools, arms, and machines flourished that lead to modern knowledge and industry. Now we begin to master energy for space travel, paving the way to new civilization.

But, what the heck, how did that happen? Energy is supposed to be consumable, so the question is, when will the sources run dry? Nobody can tell us for sure, but we hope it will not be by tomorrow!

Bibliography

What Does Energy Mean?
1. Smith, C. (1998). *The Science of Energy — A Cultural History of Energy Physics in Victorian Britain.* The University of Chicago Press.
2. Balian, R. (2013). La longue élaboration du concept d'énergie. *Académie des Sciences, 3*(09). http://www.academiesciences.fr/activite/hds/textes/evol_Balian1.pdf.
   Balian, R. (2013). *La longue élaboration du concept d'énergie.* Académie des Sciences. http://www.academiesciences.fr/activite/hds/textes/evol_Balian1.pdf. *Consulté le, 3*(09), 2013.
   https://scholar.google.com/scholar?start=10&q=roger+balian&hl=en&as_sdt=0,5&as_ylo=2013#d=gs_cit&t=1670806483984&u=%2Fscholar%3Fq%3Dinfo%3AjRDoObj4k1AJ%3Ascholar.google.com%2F%26output%3Dcite%26scirp%3D11%26hl%3Den;
   https://www.academie-sciences.fr/archivagesite/activite/hds/textes/evol-Balian1.pdf.
3. Newton, I. (1687). *Philosophiae Naturalis Principia Mathematica.* Londini, Jussu Societatis Regiæ ac Typis Josephi Streater. Prostat apud plures Bibliopolas.
4. Darwin, C. (1859). *On the Origin of Species by Means of Natural Selection, or, The Preservation of Favoured Races in the Struggle for Life.* J. Murray.
5. Kurzweil, R. (2006). *The Singularity is Near: When Humans Transcend Biology.* Penguin Books.
6. Fukuyama, F. (2003). *Our Posthuman Future: Consequences of the Biotechnology Revolution.* Farrar, Straus and Giroux.
7. Gillispie, C. C. (1981). *Dictionary of Scientific Biography.* Scribner's Sons.

8. Same as 2: Roger Ballan — a longue élaboration du concept d'énergie https://www.academie-sciences.fr/archivage_site/activite/hds/textes/evol_Balian1.pdf.

9. Bergson, H. (1920). *Mind-energy (L'Énergie spirituelle, 1919)*. McMillan.

10. *L'énergie, de quoi s'agit-il exactement ?* Jean-Marc Jancovici — Articles et études de Jean-Marc Jancovici. Ce conseiller en organisation propose services et connaissances dans les domaines de l'énergie et du climat. (2011). Retrieved December 9, 2022 from https://jancovici.com/transition-energetique/l-energie-et-nous/lenergie-de-quoi-sagit-il-exactement/.

11. Fillard, J. P. (2020). *Transhumanism: A Realistic Future?* World Scientific Publishing Company.

12. Kurzweil, R. (2006). *The Singularity is Near: When Humans Transcend Biology*. Penguin Books.

13. Integrated Fuel Cells technology — http://ifcech.com.

About the Various Forms of Energy

1. Bianconi, E., Piovesan, A., Facchin, F., Beraudi, A., Casadei, R., Frabetti, F., Vitale, L., Pelleri, M. C., Tassani, S., Piva, F., Perez-Amodio, S., Strippoli, P., & Canaider, S. (2013). An estimation of the number of cells in the human body. *Annals of Human Biology*, *40*(6), 463–471. https://doi.org/10.3109/03014460.2013.807878.

2. Neurons: Where does their electricity come from? Medical Science Navigator (2017). Retrieved December 9, 2022 from https://www.medicalsciencenavigator.com/neurons-where-does-their-electricity-come-from/.

3. Weiss, M., Sousa, F., Mrnjavac, N., *et al.* (2016). The physiology and habitat of the last universal common ancestor. *Nature Microbiology*, *1*(9), 1–8. https://doi.org/10.1038/nmicrobiol.2016.116.

4. MIT-designed project achieves major advance toward fusion energy. *MIT News*. Massachusetts Institute of Technology (2021). Retrieved December 9, 2022 from https://news.mit.edu/2021/MIT-CFS-major-advance-toward-fusion-energy-0908.

5. China maintains "artificial sun" at 120 million Celsius for over 100 seconds, setting new world record. *Global Times* (2020). Retrieved December 9, 2022 from https://www.globaltimes.cn/page/202105/1224755.shtml.

6. Asimov, I. (1997). *The Roving Mind*. Prometheus.

Discoveries follow each other in a precise order

The Historical Implementation of Energy with Tools and Machines

1. Precession of Mercury's perihelion in his 1916 paper *The Foundation of the General Theory of Relativity*.

2. Einstein, A. (1916). Die Grundlage der allgemeinen Relativitätstheorie. *Vierte Folge, 354*(7), 769–822. https://doi.org/10.1002/andp.19163540702.
3. Fitzpatrick, R. (2012). *An Introduction to Celestial Mechanics 1st Edition.* Cambridge University Press.

The Pivotal Role of Knowledge

1. Peebles, P. J. E. & Ratra, B. (2003). The cosmological constant and dark energy. *Reviews of Modern Physics, 75*(2), 559.
2. Oparin A. I. (1924). Proiskhozhozhdenie zhizny. Moscow (translated by Ann Synge, in Bernal 1967. *The Origin of Life.* Weidenfeld and Nicolson, London.
3. Oparin A. I. (1952). *The Origin of Life.* New York: Dover.
4. Maher, K. A. & Stevenson, D. J. (1988). Impact frustration of the origin of life. *Nature, 331*(6157), 612–614. https://doi.org/10.1038/331612a0.
5. Oró, J., Miller, S. L., & Lazcano, A. (1990). The origin and early evolution of life on Earth. *Annual Review of Earth and Planetary Sciences, 18*, 317–356. https://doi.org/10.1146/annurev.ea.18.050190.001533.
6. Benn, C. (2001). The Moon and the origin of life. *Earth, Moon and Planets, 85/86*(61–66). https://doi.org/10.48550/arXiv.astro-ph/0112399.

Civilizations and Induced Issues

1. Reversing Death with Stem Cells, frontlinegenomics.com.
2. Bostrom, N. (2005). A history of transhumanist thought. *Journal of Evolution and Technology, 14*(1), 1–25.
3. Fillard, J. P. (2020). *Transhumanism: A Realistic Future?* World Scientific Publishing Company.
4. Orwell, G. (1949). *1984.* Alma Classic Evergreens 2021.
5. Asimov, I. (1979). *Extraterrestrial Civilizations.* Crown.
6. Smil, V. (2017). *Energy and Civilization: A History.* The MIT Press.
7. Kardashev, N. S. (1964). Transmission of Information by Extraterrestrial Civilizations. *Soviet Astronomy, 8*, 217–221.
8. Huntington, S. (1993). The clash of civilizations? *Foreign Affairs, 72*(3), 22–49. https://doi.org/10.2307%2F20045621.
9. Transmutation of gold from mercury, Hantaro Nagaoka (1924).

Would Energy be Lasting?

1. The new fuel to come from Saudi Arabia. BBC — Homepage. (2020). Retrieved December 12, 2022 from https://www.bbc.com/future/article/20201112-the-green-hydrogen-revolution-in-renewable-energy.
2. "Tsunami of data" could consume one fifth of global electricity by 2025. Environment. *The Guardian* (2017). Retrieved December 13, 2022 from

https://www.theguardian.com/environment/2017/dec/11/tsunami-of-data-could-consume-fifth-global-electricity-by-2025.

Where Does Energy Come From?

1. Steven, W. (2013). *Physics: What We Do and Don't Know. The New York Review of Books.* Retrieved December 12, 2022 from https://www.nybooks.com/articles/2013/11/07/physics-what-we-do-and-dont-know/.
2. Hawking, S. (2018). *Brief Answers to the Big Questions.* John Murray.
3. Bondi, H. & Gold, T. (1948). The steady-state theory of the expanding universe. *Monthly Notices of the Royal Astronomical Society, 108*(3), 252–270. https://doi.org/10.1093/mnras/108.3.252.
4. Hoyle, F. (1948). A new model for the expanding universe. *Monthly Notices of the Royal Astronomical Society, 108*(5), 372–382. https://doi.org/10.1093/mnras/108.5.372.
5. https://en.wikipedia.org/wiki/Multiverse - cite_note-Richard_P._Feynman_on_The_Scientific_Method-7; https://www.youtube.com/watch?v=EYPapE-3FRw.
6. https://en.wikipedia.org/wiki/Multiverse - cite_note-67.
   Pathria, R. K. (1972). The universe as a black hole. *Nature, 240*(5379), 298–299. Bibcode:1972Natur.240.298P. doi:10.1038/240298a0. ISSN 0028-0836. S2CID 4282253.
7. https://en.wikipedia.org/wiki/Dark_matter - cite_note-15.
8. https://en.wikipedia.org/wiki/Space_exploration - cite_note-74.
   Zelenyi, L. M. Korablev, O. I. Rodionov, D. S. Novikov, B. S. Marchenkov, K. I. Andreev, O. N., & Larionov, E. V. (2015). Scientific objectives of the scientific equipment of the landing platform of the ExoMars-2018 mission. *Solar System Research, 49*(7), 509–517. Bibcode:2015SoSyR..49..509Z. doi:10.1134/S0038094615070229. ISSN 0038-0946. S2CID 124269328.
9. New telescope could reveal secrets of the universe. *Mail & Guardian — Africa's Best Read.* (2006). Retrieved December 12, 2022 from https://mg.co.za/article/2006-08-05-new-telescope-could-reveal-secrets-of-the-universe/.